AN

ESSAY

ON THE USE OF THE

Celestial and Terrestrial

GLOBES;

EXEMPLIFIED IN A GREATER VARIETY OF PROBLEMS, THAN ARE TO BE FOUND IN ANY OTHER WORK;

Exhibiting the General Principles of

DIALING AND NAVIGATION

BY THE LATE

GEORGE ADAMS,

Mathematical Instrument Maker to His Majesty, and Optician to the Prince of Wales

FIFTH EDITION.

WITH THE AUTHOR'S LAST IMPROVEMENTS,

Illustrated with Copper Plates.

PELICAN PUBLISHING COMPANY
GRETNA 2006

Copyright © 1766, 1808

First published in five editions by William Woodward
Published by Pelican Publishing Company, Inc., 2006

First edition, 1766
Fifth edition, 1808
First Pelican edition, 2006

The word "Pelican" and the depiction of a pelican are trademarks of Pelican Publishing Company, Inc., and are registered in the U.S. Patent and Trademark Office.

Printed in the United States of America
Published by Pelican Publishing Company, Inc.
1000 Burmaster Street, Gretna, Louisiana 70053

CONTENTS.

	Page.
OF the Use of the Globes	9
Advantages of Globes.	9
Description of the Globes	18
Of the Terrestrial Globe	28
Of Latitude and Longitude	28

Problem.

1. To find the Longitude of any Place	33
2. To find the difference of Longitude between any two Places	35
3. To find those places where it is Noon at any given Hour of the Day, at any given Place	36
4. When it is Noon at any Place, to find what Hour it is at any other Place	37
5. At any given Hour where you are, to find the Hour at a Place proposed	38
Of Latitude	39
6. To find the Latitude of any Place	41
7. To find all those Places which have the same Latitude with any given Places	41
8. To find the Difference of Latitude between any two Places	42

CONTENTS.

Problem.	Page.
9. The Latitude and Longitude being known, to find the Place	42
Of finding the Longitude	43
10. To find the Distance of one Place from another	53
11. To find the Angle of Position of Places	54
12. To find the Bearings of Places	54
Of the twilight	45
To rectify the Globe	59
13. To rectify for the Summer Solstice	61
14. ——— for the Winter Solstice	63
15. ——— for the Times of Equinox	64
16. To exemplify the Sun's Altitude	67
17. Of the Sun's Meridian Altitude	68
18. To find the Sun's Meridian Altitude universally	69
19. Of the Sun's Azimuths	70
Of the Zones and Climates	72
20. To find the Climates	74
21. To illustrate the Distinction of Ascii, &c.	78
22. To find the Antœci, &c.	81
23. To find those Places over which the Sun is verticial	82
24. To find the Sun's Place	83
25. To find the Sun's Declination	86
26. To find the two Days on which the Sun is in the Zenith of any given Place, &c.	87
27. To find where the Sun is vertical on a given Day and Hour	87
28. At a given Time of the Day in one Place, to find at the same Instant those Places where the Sun is rising, setting, &c.	88
29. To find all those Places within the Polar Circles, on which the Sun begins to shine, &c.	90
30. To make Use of the Globe as a Tellurian	91
31. To rectify the Globe to the Latitude and Horizon of any Place	95
32. To rectify for the Sun's Place	95

CONTENTS.

Problem.	Page.
33 To rectify for the Zenith of any Place	96
Of exposing the Globe to the Sun	97
34. To observe the Sun's Altitude	100
35. To place the Globe, when exposed to the Sun, that it may represent the natural Positions of the Earth	102
36. To find naturally the Sun's Declination	104
37. To find naturally the Sun's Azimuth	105
38. To shew where the Sun will be twice on the same Azimuth in the Morning, and twice in the Afternoon	106
To find the Hour by the Sun	108
Of Dialling	112
40. To construct an Horizontal Dial	117
41. To delineate a South Dial	121
42. To make an erect Dial	122
Of Navigation	126
43. Given the Difference of Latitude, and Difference of Longitude, to find the Course and Distance sailed	132
44. Given the Difference of Latitude and Course, to find the Difference of Longitude and Distance sailed	133
45. Given the Difference of Latitude and Distance run, to find the Difference of Longitude, and Angle of the Course	134
46. Given the difference of Longitude and Course, to find the difference of Latitude, and Distance sailed	135
47. Given the Course and Distance, to find the Difference of Longitude and Latitude	136
48. To steer a ship upon the Arch of a great Circle, &c.	137
Of the Celestial Globe	151
Of the Precession of the Equinoxes	157

CONTENTS.

Problem.	Page.
2. To rectify the Celestial Globe	163
3. To find the Declination and Right Ascension of the Sun	164
4. To find the Sun's oblique Ascension, &c.	165
5. ——— the Sun's meridian Altitude	166
6. ——— the Length of the Day in Latitudes under 66½ Degrees	166
7. ——— the Length of the longest and shortest Day in Latitudes under 66½ Degrees	167
8. To find the Latitude where the longest Day may be of any given Length between twelve and twenty four Hours	167
9. ——— the time of Sun-rising, &c.	168
10. ——— how long, &c. the Sun shines in any Place within the Polar Circles	170
11. To illustrate the Equation of Time, &c.	174
12. To find the Right Ascension, &c of a Star	176
13. ——— the Latitude and Longitude of a Star	177
14. ——— the Place of a Star on the Globe by, &c.	177
15. ——— at what hour a given Star transits the meridian	178
16. On what Day a Star will come to the Meridian	179
17. To represent the Face of the Heavens for any given Day and Hour	179
18. To trace the Circles of the sphere in the Heavens	182
19. To find the Circle of perpetual Apparition	188
20. ——— the Sun's Amplitude	189
21. ——— the Sun's Altitude at a given Hour	190
22. ——— when the Sun is due East in a given Latitude	193
23. ——— the Rising, Setting, Culminating, &c. of a Star	194
24. ——— the Hour of the Day, the Altitude and Azimuth of a Star being given	195
25. ——— the Altitude and Azimuth of a Star, &c.	196

CONTENTS. vii

Problem.		Page.
26. —— the Azimuth, &c. at any Hour of the Night		197
27. —— the Sun's Altitude, and the Hour, from the Latitude, Sun's Place, and Azimuth		197
21. —— the Hour, the Latitude and Azimuth given		198
29. —— a Star, the Latitude, Sun's Place, Hour, &c. given		199
30. To find the Hour by Data from two Stars that have the same Azimuth		199
31. —— the Hour by Data from two Stars that have the same Altitude		200
32. —— the Latitude by Data from two Stars		201
33. —— the Latitude by other Data from two Stars		201
34. —— when a Star rises or set cosmically		203
35. —— when a Star rises or sets achronically		204
36. —— when a Star will rise heliacally		206
37. —— when a Star will set heliacally		207
Of the Correspondence between the Celestial and Terrestrial Spheres		208
28. To find the Place of a Planet, &c.		212
39. —— what Planets are above the Horizon		213
40. —— the right Ascension, &c. of a Planet		214
41. —— the Moon's Place		220
42. —— the Moon's Declination		221
43. —— The Moon's greatest and least Meridian Altitudes		222
44. To illustrate the Harvest Moon		223
45. To find the Azimuth of the Moon, and thence High Water, &c.		228
Of Comets		229
46. To rectify the Globe for the Place of Observation		231
47. To determine the Place of a Comet		232
48. To find the Latitude, &c. of a Comet		232
49. To find the Time of a Comet's Rising, &c.		233

CONTENTS.

Problem.	Page.
50. To find the same at London	234
51. To determine the Place of a Comet from an Observation made at London	234
52. From two given Places to assign the Comet's Path	235
53. To estimate the Velocity of a Comet	236
54. To represent the general Phenomena of a Comet	237

PREFACE

TO THE ESSAY ON THE GLOBES.

THE connection of astronomy with geography is so evident, and both in conjunction so necessary to a liberal education, that no man will be thought to have deserved ill of the republic of letters, who has applied his endeavours to diffuse more universally the knowledge of these useful Sciences, or to render the attainment of them easier; for as no branch of literature can be fully comprehended without them, so there is none which impresses more pleasing ideas on the mind, or that affords it a more rational entertainment.

In the present work, several objections to former editions are obviated; the Problems arranged in a more methodical manner, and a great number added. Such facts are also occasionally introduced, such observations interspersed, and such relative information communicated, as it is presumed will excite curiosity, and fit attention.

To further the design, the attention is directed to the appearance of the planetary bodies, as observed from the earth. It were to be wished that the tutor would at this part exhibit to his pupil the various phenomena in the heavens themselves; by teaching him thus to observe for himself, he would not only raise his curiosity, but so fix the impressions which the objects have made on his mind, that by proper cultivation they would prove a fruitful source of useful employment; and he would therby also gratify that eager desire after novelty, which continually animates young minds, and furnishes them with objects on which to exercise their natural activity.

PART I.

A TREATISE

ON THE USE OF THE TERRESTRIAL AND CELESTIAL GLOBES.

OF THE ADVANTAGES OF GLOBES IN GENERAL, FOR ILLUSTRATING THE PRIMARY PRINCIPLES OF ASTRONOMY AND GEOGRAPHY; AND PARTICULARLY OF THE ADVANTAGES OF THE GLOBES, WHEN MOUNTED IN MY FATHER'S MANNER.

UNIVERSAL approbation, the opinion of those that excel in science, and the experience of those that are learning, all concur to prove that the artificial representations of the earth and heavens, on the terrestrial and celestial globes, are the instruments the best adapted to convey natural and genuine ideas of astronomy and geography to young minds.

This superiority they derive principally from their form and figure, which communicates a more just idea, and gives a more adequate representation of the earth and heavens, than can be formed from any other figure.

To understand the nature of the projection of either sphere in plano, requires more knowledge of geometry than is generally possessed by beginners, it's principles are more recluse, and the solution of problems more obscure.

The motion of the earth upon it's axis is one of the most important principles both in geography and astronomy; on it the greater part of the phenomena of the visible world depend: but there is no invention that can communicate so natural a representation of this motion, as that of a terrestrial globe about it's axis. By a celestial globe, the apparent motion of the heavens is also represented in a natural and satisfactory manner.

In order to convey a clear idea of the various divisions of the earth, of the situation of different places, and to obtain an easy solution of the various problems in geography, it is necessary to conceive many imaginary circles delineated on it's surface, and to understand their relation to each other. Now on a globe these circles have their true form; their intersections and relative positions are visible upon the most cursory inspection. But in projections of the sphere in plano, the form of these circles is varied, and their nature changed; they are consequently but ill adapted to convey to young minds the elementary principles of geography.

On a globe, the appearance of the land and water is perfectly natural and continuous, fitted to convey accurate ideas, and leave permanent impressions on the most tender minds; whereas in planispheres one-half of the globe is separated and disjoined from the other; and those parts, which are contiguous on a globe, are here separated and thrown at a distance from each other. The celestial globe has the same superiority over projections of the heavens in plano.

The globe exhibits every thing in true proportion, both of figure and size; while on a planisphere the reverse may often be observed.

Presuming that these reasons sufficiently evince the great advantage of globes over either planispheres or maps, for obtaining the first principles of astronomical and geographical knowledge, I proceed to point out the pre-eminence of globes *mounted in my father's manner*, over the common, or rather the old and Ptolemaic mode of fitting them up.

The great and increasing sale of his globes mounted in the best manner, may be looked upon at least as a proof of approbation from numbers; to this I might also add, the encouragement they have received from the principal tutors of both our universities, the public sanction of the university of Leyden, the many editions of my father's treatise on their use, and its translation into Dutch, &c. The recommendation of Mess. Arden, Walker, Burton, &c. public lecturers in natural philosophy, might also be adduced: but leaving these considerations, I shall proceed to enumerate the reasons which give them, in my opinion, a decided preference over every other kind of mounting.*

* The following note from Mr. Walker's Easy Introduction to Geography, in favour of my father's globes, will not, I hope, be deemed improper.

" Simplicity and perspicuity should ever be studied by those who cultivate the young mind; and jarring, opposing, or equivocal ideas should be avoided almost as much as error or falsehood. Our globes, till of late years, were equipt with an hour circle, which prevented the poles from sliding through the horizon; hence their rectification was generally for the *place on the earth*, instead of the *sun's place in the ecliptic;* which put the globe into so unnatural and absurd a position respecting the sun, that young people were confounded when they compared it with the earth's positions during it's annual rotation round that luminary, and considering the horizon as the boundary of day and night. Being, therefore, sometimes obliged to rectify for the place on the earth, and sometimes for the sun's place in the ecliptic, the two rules clash so unhappily in the pupil's mind, that few remember a single problem a twelvemonth after the end of their tuition. Globes, therefore, with a horary circle, are but partially described in this treatise; the great intention of which is, to make the elevations and

The earth, by it's diurnal revolution on it's axis, is carried round from west to east. To represent this real motion of the earth, and to solve problems agreeable thereto, it is necessary that the globe, in the solution of every problem, should be moved from *west to east;* and for this purpose, that the divisions on the large brass circle should be on that side which looks westward.* Now this is the case in my father's mode of mounting the globes, and the tutor can thereby explain with ease the rationale of any problem to his pupil. But in the common mode of mounting, the globe must be moved from east to west, according to the Ptolemaic system; and consequently, if the tutor endeavours to shew how things obtain in nature, he must make his pupil unlearn in a degree what he has taught him, and by abstraction reverse the method he has instructed him to use; a practice that we hope will not be adopted by many.

depressions of the poles of a terrestrial globe to represent *all* the situations the earth is in to the sun, for every day or hour through the year. The globes of *Mr. Adams* are the most favourable for the above mode of rectification of any plates we have at present; and to make a quiescent globe to represent all the positions of one revolving round the sun, turning on an inclined axis, and keeping that axis altogether parallel to itself, his globes are better adapted than any, I believe, in being."

* See the Rev. Mr. Hutchin's New Treatise on the Globes.

The celestial globe being intended to represent the apparent motion of the heavens, should be moved, when used, from east to west.

Of the phenomena to be explained by the terrestrial globe, the most material are those which relate to the changes in the seasons; all the problems connected with, or depending upon these phenomena, are explained in a clear, familiar, and natural manner, by the globe, when mounted in my father's mode; for on rectifying it for any particular day of the month, it immediately exhibits to the pupil the exact situation of the globe of the earth for that day; and while he is solving his problem, the reason and foundation of it presents itself to the eye and understanding.

The globe may also be placed with ease in the position of a right sphere; a circumstance exceedingly useful, and which the old construction of the globes did not admit of.

By the application of a moveable meridian, and an artificial horizon connected with it, it is easy to explain why the sun, although he be always in one and the same place, appears to the inhabitants of the earth at different altitudes, and in different azimuths, which cannot be so readily done with the common globes.

On the celestial globe there is a moveable circle of declination, with an artificial sun.

The brass wires placed under the globes, serve to distinguish, in a natural and satisfactory manner, twilight from total darkness, and the reason of the length of it's duration.

The next point, wherein they materially differ from other globes, is in the hour circle. Now it must be confessed, that to every contrivance that has been used for this purpose there is some objection, and probably no mode can be hit upon that will be perfectly free from them. The method adopted by my father appears to me the least exceptionable, and to possess some advantages over every other method I am acquainted with. Agreeably to the opinion of the first astronomers, among others of M. de la Lande, he uses the equator for the hour circle, not only as the largest, but also as the most natural circle that could be employed for that purpose, and by which alone the solution of problems could be obtained with the greatest accuracy. As on the terrestrial globe, the longitude of different places is reckoned on this circle; and on the celestial, the right ascension of the stars, &c. it familiarizes the young pupil with them, and their reduction to time. This method does not in the least impede the motion of the globe; but while it affords an equal facility of elevating either the north or south pole, it prevents the pupil from placing them in a wrong position; while the horary wire secures the globe from falling out of the frame.

Another circumstance peculiar to these globes, is the mode of fixing the compass. It is self-evident, that the tutor, who is willing to give correct ideas to his pupil, should always make him keep the globes with the north pole directed towards the north pole of the heavens, and that, both in the solution of problems, and the explanation of phenomena. By means of the compass, the terrestrial globe is made to supply the purpose of a tellurian, when such an instrument is not at hand. I cannot terminate this paragraph, without testifying my disapprobation of a mode adopted by some, of making the globe turn round upon a pin in the pillar on which it is supported; a mode, that, while it can give little but relief to indolence, is less firm in it's construction, and tends to introduce much confusion in the mind of the pupil.

In order to prevent that confusion and perplexity which necessarily arises in a young mind, when names are made use of which do not properly characterize the subject, my father found it necessary, with Mr. Hutchins, to term that broad wooden circle which supports the globe, and on which the signs of the ecliptic and the days of the month are engraved, the *broad paper circle,* instead of horizon, by which it had been heretofore denominated. The propriety of this change will be evident to all those who consider, that this circle in some cases represents that which divides light from darkness, in others the horizon, and sometimes the ecliptic. For similar reasons, he was induced to call the brazen circle, in which the globes are suspended, the *strong brass circle*.

In a word, many operations may be performed by these globes, which cannot be solved by those mounted in the common manner; while all that they can solve may be performed by these, and that with a greater degree of perspicuity; and many problems may be performed by these at one view, which on the other globes require successive operations.

But as, notwithstanding their superiority, the difference in price may make some persons prefer the old construction, it may be proper to inform them, that they may have my *father's* globes mounted in the *old manner*, at the usual prices.

PART II.

CONTAINING

A DESCRIPTION OF THE GLOBES MOUNTED IN THE BEST MANNER; TOGETHER WITH SOME PRELIMINARY DEFINITIONS.

DEFINITIONS.

BEFORE we begin to discribe the globes, it will be proper to take some notice of the properties of a circle, of which a globe may be said to be constituted.

A *line* is generated by the motion of a point.

Let there be supposed two points, the one moveable, the other fixed.

If the moveable point be made to move directly towards the fixed point, it will generate in it's motion a straight line.

If a moveable point be carried round a fixed point, keeping always the same distance from it, it will generate a *circle*, or some part

OF THE GLOBES.

of a circle, and the fixed point will be the *center* of that circle.

All strait lines going from the center to the circumference of a circle, are equal.

Every strait line that passes through the center of a globe, and is terminated at both ends by it's surface, is called a *diameter*.

The extremities of a diameter are it's poles.

If the circumference of a semicircle be turned round it's diameter, as on an axis, it will generate a globe, or sphere.

The center of the semicircle will be the center of the globe; and as all points of the generating semicircle are at an equal distance from it's center, so all the points of the surface of the generated sphere are at an equal distance from it's center.

DESCRIPTION OF THE GLOBES.

There are two artificial globes. On the surface of one of them the heavens are delineated; this is called the *celestial globe*. The other, on which the surface of the earth is described, is called the *terrestrial globe*.

Fig. 2, plate XIII, represents the celestial, fig. 1, plate XIII, the terrestrial globe, as mounted in my father's manner.

In using the celestial globe, we are to consider ourselves as at the *center*.

In using the terrestrial globe, we are to suppose ourselves on some point of it's *surface*.

The motion of the terrestrial globe represents the *real* motion of the earth.

The motion of the celestial globe represents the *apparent* motion of the heavens.

The motion, therefore, of the celestial globe, is a motion from *east to west*.

But the motion of the terrestrial globe is a motion from *west to east*.

On the surface of each globe several circles are described, to every one of which may be applied what has been said of circles in page 205.

The center of some of these circles is the same with the center of the globe; these are, by way of distinction, called *great circles*.

Of these great circles, some are graduated.

The graduated circles are divided into 360, or equal parts, 90 of which make a quarter of a circle, or a quadrant.

Those circles, whose centers do not pass through the center of the globe, are called *lesser circles*.

The globes are each of them suspended at the poles in a strong brass circle N Z Æ S, and turn therein upon two iron pins, which are the axis of the globe; they have each a thin brass semicircle N H S, moveable about these poles, with a small thin circle H sliding thereon: it is quadrated each way to 90° from the equator to either pole.

On the terrestrial globe this semicircle is a *moveable meridian.* It's small sliding circle, which is divided into a few of the points of the mariner's compass, is called a *terrestrial* or *visible horizon.*

On the celestial globe this semicircle is a *moveable circle of declination,* and it's small brass circle an artificial sun, or planet.

Each globe has a brass wire circle, T W Y, placed at the limits of the crepusculum, or twilight, which, together with the globe, is mounted in a wooden frame. The upper part, B C, is covered with a broad paper circle, whose plane divides the globe into two hemispheres; and the whole is supported by a neat pillar and claw, with a magnetic needle in a compass-box, marked M.

A DESCRIPTION OF THE CIRCLES DESCRIBED ON THE BROAD PAPER CIRCLES B C; TOGETHER WITH A GENERAL ACCOUNT OF IT'S USES.

It contains four concentric circular spaces, the innermost of which is divided into 360°.

and numbered into four quadrants, beginning at the east and west points, and proceeding each way to 90°, at the north and south points: these are the four cardinal points of the horizon. The second circular space contains, at equal distances, the thirty-two points of the mariner's compass. Another circular space is divided into twelve equal parts, representing the twelve signs of the zodiac; these are again subdivided into 30 degrees each, between which are engraved their names and characters. This space is connected with a fourth, which contains the calendar of the months and days; each day, on the eighteen-inch globes being divided into four parts, expressing the four cardinal points of the day, according to the Julian reckoning; by which means the sun's place is very nearly obtained for the common years after bissextile, and the intercalary day is inserted without confusion.

In all positions of the celestial globe, this broad paper circle represents the plane of the horizon, and distinguishes the visible from the invisible part of the heavens; *but in the terrestrial globe, it is applied to three different uses.*

1. To distinguish the points of the horizon. In this case it represents the *rational* horizon of any place.

2. It is used to represent the circle of

illumination, or that circle which separates day from night.

3. It occasionally represents the *ecliptic*.

Of the strong brass circle N Æ Z S. One side of this strong brass circle is graduated into four quadrants, each containing 90 degrees.

The numbers on two of these quadrants increase from the equator towards the poles; the other two increase from the poles towards the equator.

Two of the quadrants are numbered from the equator, to shew the distance of any point on the globe from the equator. The other two are numbered from the poles, for the more ready setting the globe to the latitude of any place.

The strong brass circle of the celestial globe is called the meridian, because the centre of the sun comes directly under it at noon.

But as there are other circles on the terrestrial globe, which are called meridians, we chuse to denominate this the *strong brass circle*, or *meridian*.

The graduated side of the strong brass circle, that belongs to the terrestrial globe, should face the *west*.

The graduated side of the strong brazen meridian of the celestial globe, should face the *east*.

On the strong brass circle of the terrestrial globe, and at about 23½ degrees on each side of the north pole, the days of each month are laid down according to the declination of the sun.

Of the Horary Circles, and their Indices. When the globes are mounted in my father's manner, we use the equator as the hour circle; because it is not only the most natural, but also the largest circle that can be applied for that purpose.

To make this circle answer the purpose, a semi-circular wire is placed over it, carrying two indices, one on the east, the other on the west side of the strong brass circle.

As the equator is divided into 360°, or 24 hours, the time of one entire revolution of the earth or heavens, the indices will shew in what space of time any part of such revolution is made among the hours which are graduated below the degrees of the equator on either globe.

As the motion of the terrestrial globe is from west to east, the horary numbers increase according to the direction of that motion: on the celestial globe they increase from the east to the west.

Of the Quadrant of Altitude, Z A. This is a thin, narrow, flexible slip of brass, that will bend to the surface of the globe; it has a nut, with a fiducial line upon it, which may be readily applied to the divisions on the strong brass meridian of either globe. One edge of the quadrant is divided into 90 degrees, and the divisions are continued to 18 degrees below the horizon.

OF SOME OF THE CIRCLES THAT ARE DESCRIBED UPON THE SURFACE OF EACH GLOBE.

We may suppose as many circles to be described on the surface of the earth as we please, and conceive them to be extended to the sphere of the heavens, making thereon concentric circles: for as we are obliged, in order to distinguish one place from another, to appropriate names to them, so are we obliged to use different circles on the globes, to distinguish their parts, and their several relations to each other.

Of the Equator, or Equinoctial. This circle goes round the globe exactly in the middle, between the two poles, from which it always keeps at the same distance; or in other words, it is every where 90 degrees distant from each pole, and is therefore a boundary, separating the northern from the southern hemisphere; hence it is frequently called *the line* by sailors, and when they sail over it they are said to cross the line.

It is that circle in the heavens in which the sun appears to move on those two days, the one in the spring, the other in the autumn, when the days and nights are of an equal length all over the world; and hence on the celestial globe it is generally called the *equinoctial*.

It is graduated into 360 degrees. Upon the terrestrial globe the numbers increase from the meridian of London westward, and proceed quite round to 360. They are also numbered from the same meridian eastward, by an upper row of figures, to accomodate those who use the English tables of latitude and longitude.

On the celestial globe, the equatorial degrees are numbered from the first point of Aries eastward, to 360 degrees.

Under the degrees on either globe is graduated a circle of hours and minutes. On the celestial globe the hours increase eastward, from Aries to XII at Libra, where they begin again in the same direction, and proceed to XII at Aries. But on the terrestrial globe, the horary numbers increase by twice twelve hours westward from the meridian of London to the same again.

In turning the globe about, the equator keeps always under one point of the strong brass meridian, from which point the degrees on the said circle are numbered both ways.

Of the Ecliptic. The graduated circle, which crosses the equator obliquely, forming with it an angle of about 23½ degrees, is called the ecliptic.

This circle is divided into twelve equal parts, each of which contains thirty degrees. The beginning of each of these thirty degrees is marked with the characters of the twelve signs of the zodiac.

The sun appears always in this circle; he advances therein every day nearly a degree, and goes through it exactly in a year.

The points where this circle crosses the equator are called the *equinoctial points*. The one is at the beginning of Aries, the other at the beginning of Libra.

The commencement of Cancer and Capricorn are called the *solstitial points*.

The twelve signs, and their degrees, are laid down on the terrestrial globe; but upon the celestial globe, the days of each month are graduated just under the ecliptic.

The ecliptic belongs principally to the celestial globe.

PART III.

THE USE OF THE TERRESTRIAL GLOBE, MOUNTED IN THE BEST MANNER.

OF LONGITUDE AND LATITUDE, OF TERRESTRIAL MERIDIANS, AND THE PROBLEMS RELATING TO LONGITUDE AND LATITUDE.

MERIDIANS are circular lines, going over the earth's surface, from one pole to the other, and crossing the equator at right angles.

Whatever places these circular lines pass through, in going from pole to pole, they are the meridians of those places.

There are no places upon the surface of the earth, through which meridians may not be conceived to pass. Every place, therefore, is supposed to have a meridian line passing over it's zenith from north to south, and going through the poles of the world.

Thus the meridian of Paris is one meridian; the meridian of London is another. This variety of meridians is satisfactorily represented on the globe, by the moveable meridian, which may be set to every individual point of the equator, and put directly over any particular place.

Whensoever we move towards the east or west, we change our meridian; but we do not change our meridian if we move directly to the north or south.

The moveable meridian shews that the poles of the earth divide every meridian into two semicircles, one of which passes through the place whose meridian it is, the other through a point on the earth, opposite to that place.

Hence it is, that writers in geography and astronomy generally mean by the *meridian* of any place the *semicircle* which passes through that place; these, therefore, may be called the geographical meridians.

All places lying under the same semicircle, are said to have the same meridian; and the semicircle opposite to it is called the opposite meridian, or sometimes the opposite part of the meridian.

From the foregoing definitions, it is clear that the meridian of any place is immoveably fixed to that place, and is carried round along with it by the rotation of the globe.

When the meridian of any place is by the revolution of the earth brought to point at the sun, it is noon, or mid-day, at that place.

The plane of the meridian of any place may be imagined to be extended to the sphere of the fixed stars.

When, by the motion of the earth, the plane of a meridian comes to any point in the heavens, as the sun, moon, &c. that point, &c. is then said to come to the meridian. It is in this sense that we generally use the expression of the sun or stars coming to, or passing over the meridian.

The time which elapses between the noon of any one day in a given place, and the noon of the day following in the same place, is called *a natural day.*

All places which lie under the same meridian, have their noon, and every other hour of the natural day, at the same time. Thus when it is one in the afternoon at London, it is also one in the afternoon to every place under the meridian of London.

In order to ascertain the situation of any point, there must first be a settled part of the earth's surface, from which to measure; and as the point to be ascertained may lie in any part of the earth's surface, and as this surface is spherical, the place from whence we measure must be a circle. It would be necessary, however to establish two such circles; one to know how far any place may be east or west of another, the second to know it's distance north or south of the given point, and thus determine it's precise situation.

Hence it has been customary for geographers to fix upon the meridian of some remarkable place, *as a first meridian, or standard;* and to reckon the distance of any place to the east or west, or it's longitude, by it's distance from the first meridian. On English globes, this first meridian is made to pass through London. The position of this first meridian is arbitrary, because on a globe, properly speaking, there is neither beginning nor end. The first person (whose works at least are come down to us) who computed the distance of places by longitudes and latitudes was Ptolemy, about the year after Christ 140.

The *longitude of any place* is it's distance from the first meridian, measured by degrees on the equator.

To find the longitude of a place, is to find what degree on the equator the meridian of that place crosses.

All places that lie under the same meridian, are said to have the same longitude; all places that lie under different meridians, are said to have different longitudes; this difference may be east or west, and consequently the difference of longitude between any two places, is the distance of their meridians from each other measured on the equator.

Thus if the meridian of any place cuts the equator in a point, which is fifteen degrees east from that point, where the meridian of London cuts the equator, that place is said to differ from London in longitude 15 degrees eastward.

Upon the terrestrial globe there are 24 meridians, dividing the equator into 24 equal parts, which are the hour circles of the places through which they pass.

The distance of these meridians from each other is 15 degrees, or the 24th part of 360 degrees; thus 15 degrees is equal to one hour.

By the rotation of the earth, the plane of every meridian points at the sun, one hour after that meridian which is next to it eastward; and thus they successively point at the sun every hour, so that the planes of the 24 meridian semicircles being extended, pass through the sun in a natural day.

To illustrate this, suppose the plane of the strong brass meridian to coincide with the sun, bring London to this meridian, and then move the globe round, and you will find these 24 meridians successively pass under the strong brass meridian, at one hour's distance from each other; till in 24 hours the earth will return to the same situation, and the meridian of

OF THE GLOBES.   33

London will again coincide with the strong brass circle.

By passing the globe round, as in the foregoing article, it will be evident to the pupil, that if one of these meridians, 15 degrees east of London, comes to the strong brass meridian, or points at the sun one hour sooner than the meridian of London, a meridian that is 30 degrees east comes two hours sooner, and so on; and consequently they will have noon, and every other hour, so much sooner than at London: while those, whose meridian is 15 degrees westward from London, will have noon and every other hour of the day, one hour later than at London, and so on, in proportion to the difference of longitude. These definitions being well understood, the pupil will be prepared not only to solve, but see the rationale of the following problems.

PROBLEM I.

To find the Longitude of any place on the Globe.

The reader will find no difficulty in solving this problem, if he recollects the definition we have given of the word longitude, namely, that it is the distance of any place from the first meridian measured on the equator. Therefore, either set the moveable meridian to the place, or bring the place under the strong brass meridian, and that degree of the equator, which is cut by either of the brazen meridians, is the longitude in degrees and minutes, or the hour and minute of its longitude, expressed in time.

As the given place may lie either east or west of the first meridian, the longitude may be expressed accordingly.

It appears most natural to reckon the longitude always westward from the first meridian; but it is customary to reckon one half round the globe eastward, the other half westward from the first meridian. To accomodate those who may prefer either of these plans, there are two sets of numbers on our globes: the numbers nearest the equator increase westward, from the meridian of London quite round the globe to 360°, over which another set of numbers is engraved, which increase the contrary way; so that the longitude may be reckoned upon the equator, either east or west.

Example. Bring Boston, in New England, to the graduated edge of either the strong brass, or of the moveable meridian, and you will find it's longitude in degrees to be $70\frac{1}{2}$, or 4 h. 42 min. in time; Rome $12\frac{1}{2}$ degrees east, or 50 min. in time; Charles-Town, North-America, is 79 deg. 50 min. west.

PROBLEM II.

To find the difference of longitude between any two places.

If the pupil understands what is meant by the difference of longitude, the rule for the solution of this problem will naturally occur to his mind. Now the difference of longitude between any two places is the quantity of an angle (at the pole) made by the meridians of those places measured on the equator. To express this angle upon the globe, bring the moveable meridian to one of the places, and the other place under the strong brass circle, and the required angle is contained between these two meridians, the measure or quantity of which is to be counted on the equator.

Example. I find the longitude of Rome to be $12\frac{1}{2}$ east, that of Constantinople to be 29; the difference is $17\frac{1}{2}$ degrees. Again, I find Jerusalem has 35 deg. 25 min. east longitude from London; and Pekin, in China, 116 deg. 52 min. east longitude; the difference is 81 deg. 27 min.; that is, Pekin is 81 deg. 27 min. east longitude from Jerusalem; or Jerusalem is 81 deg. 27 min. west longitude from Pekin.

If one place is east, and the other west of the first meridian, either find the longitude of both places westward, by that set of numbers which increase westward from the meridian of London to 360 deg. and the difference between the number thus found is the answer to the question:—or, add the east and west longitudes, and the sum is the difference of longitude; thus the longitude of Rome is 12 deg. 30 min. east, of Charles-Town 79 deg. 50 min, west; their sum, 91 deg. 20 min. is the difference required.

It may be proper to observe here, that *the difference of time is the same with the difference of longitude*, consequently that some of the following problems are only particular cases of this problem, or readier modes of computing this difference.

PROBLEM III.

To find all those places where it is noon, at any given hour of the day, at any given place.

General rule. Bring the given place to the brass meridian; and set the index to the uppermost XII; then turn the globe, till the index points to the given hour, and it will be noon to all the places under the meridian.

As the diurnal motion of the earth is from *west to east*, it is plain that all places which are to the east of any meridian, must necessarily pass by the sun before a meridian which is to the west can arrive at it.

N. B. As in my *father's* globes, the XII, or first meridian, passes through London, you have only to bring the given hour to the east of London, if in the morning, to the brass meridian, and all those places which are under it will have noon at the given hour; but bring the given hour westward of London, if it be in the afternoon.

When it is 4 h. 50 min. in the afternoon at Paris, it is noon at New Britain, New England, St. Domingo, Terra Firma, Peru, Chili, and Terra del Fuego.

When it is 7 h. 50 min. in the morning at Ispahan, it is noon at the middle of Siberia, Chinese Tartary, China, Borneo.

PROBLEM IV.

When it is noon at any place, to find what hour of the day it is at any other place.

Rule. Bring the place at which it is noon, to the strong brass meridian, and set the hour index to the uppermost XII, and then turn the globe about till the other place comes under the strong brass meridian, and the hour index will shew upon the equator the required hour. If to the eastward of the place where it is noon, the hour found will be in the afternoon; if to the westward, it will be in the forenoon.

Thus when it is noon at London, it is 50 min. past XII, at Rome; 32 min. past VII in the evening at Canton, in China; 15 min. past VII in the morning at Quebec, in Canada.

PROBLEM V.

The hour being given at any place, to tell what hour it is in any other part of the world.

Rule. Bring the place where the time is required under the strong brass meridian, set the hour index to the given time, then turn the globe, till the other place is under the brass meridian, and the horary index will point to the hour required.

Thus suppose we are at London at IX o'clock in the morning, what is the time at Canton, in China? Answer, 31 min. past IV in the afternoon. When it is IX in the evening at London, it is about 15 min. past IV in the afternoon at Quebec in Canada.

Thus also when it is III in the afternoon at London, it is 18 min. past X in the forenoon at Boston. When it is VI in the morning at the Cape of Good Hope, it is 7 min. after midnight at Quebec.

OF LATITUDE.

I have already observed, that the equator divides the globe into two hemispheres, the northern and the southern.

The latitude of a place is it's distance from the equator towards the north or south pole, measured by degrees upon the meridian of the place.

All places, therefore, that lie under the equator, are said to have *no latitude.*

All other places upon the earth are said to be in north or south latitude, as they are situated on the north or south side of the equator; and the latitude of any place will be greater or less, according as it is farther from, or nearer to the equator.

Lines, which keep always at the same distance from each other, are called *parallels.*

If a circle, or circular line, be conceived keeping at the same distance from the equator, it will be a parallel to the equator.

Circles of this kind are commonly drawn on the terrestrial globe, on both sides of the equator.

A circle of this kind, at 10 degrees from the equator, is called a parallel of 10 degrees.

When any such parallel passes through two

places on the globe's surface, those two places have the same latitude.

Hence parallels to the equator are called *parallels of latitude*.

There are four principal lesser circles parallel to the equator, which divide the globe into five unequal parts, called *zones*.

The circle on the north side of the equator is called the *tropic of Cancer;* it just touches the north part of the ecliptic, and shews the path the sun appears to describe, the longest day in summer.

That which is on the south side of the equator is called the *tropic of Capricorn;* it just touches the south part of the ecliptic, and shews the path the sun appears to describe, the shortest day in winter.

The space between these two tropics, which contains about 47 degrees, was called by the ancients the *torrid zone.*

The two polar circles are placed at the same distance from the poles, that the two tropics are from the equator.

One of these is called the *northern*, the other the *southern polar circle.*

These include $23\frac{1}{2}$ degrees on each side of their respective poles, and consequently contain 47 degrees, equal to the number of degrees included between the tropics.

The space contained within the northern

polar circle, was by the ancients called the *north frigid zone;* and that within the southern polar circle, the *south frigid zone.*

The spaces between either polar circle, and its nearest tropic, which contain about 43 degrees each, were called by the ancients the *two temperate zones.*

PROBLEM VI.

To find the latitude of any place.

If the pupil comprehends the foregoing definition, he will find no difficulty in the solution of this and some of the following problems.

Rule. Bring the place to the graduated side of the strong brass meridian, and the degree which is over it is the latitude. Thus London will be found to have 51 deg. 30 min. north latitude; Constantinople 41 deg. north latitude; and the Cape of Good Hope 34 deg. south latitude.

PROBLEM VII.

To find all those places which have the same latitude with any given place.

Suppose the given place to be London; turn the globe round, and all those places which pass under the same point of the strong brass meridian, are in the same latitude.

PROBLEM VIII.

To find the difference of latitude between two places.

Rule. If the places be in the same hemisphere, bring each of them to the meridian, and subtract the latitude of one from the other. If they are in different hemispheres, add the latitude of one to that of the other.

Example. The latitude of London is 51 deg. 32 min.; that of Constantinople 41 deg.; their difference is 10 deg. 32 min. The difference between London, 51 deg. 32 min. north, and the Cape of Good Hope, 34 deg. south, is 84 deg. 32 min.

PROBLEM IX.

The latitude and longitude of any place being known, to find that place upon the globe.

Rule. Seek for the given longitude in the equator, and bring the moveable meridian to that point; then count from the equator on the meridian, the degree of latitude either towards the north or south pole, and bring the artificial horizon to that degree, and the intersection of it's edge with the meridian is the situation required.

By this problem any place not represented on the globe may be laid down thereon, and it may be seen where a ship is when it's latitude and longitude are known.

Example. The latitude of Smyrna, in Asia, is 38 deg. 28 min. north; it's longitude 27 deg. 30 min. east of London; therefore, bring 27 deg. 30 min. counted eastward on the equator, to the moveable meridian, and slide the diameter of the artificial horizon to 38 deg. 28 min. north latitude, and it's center will be correctly placed over Smyrna.

It may be proper in this place just to shew the pupil, that *the latitude of any place is always equal to the elevation of the pole of the same place above the horizon.* The reason of this is, that from the equator to the pole are 90 degrees, from the zenith to the horizon are also 90 degrees; the distance of the zenith to the pole is common to both, and therefore if taken away from both, must leave equal remains; that is, the distance from the equator to the zenith, which is the latitude, is equal to the elevation of the pole.

OF FINDING THE LONGITUDE.

As the finding the longitude of places forms one of the most important problems in geography and astronomy, some further account of it, it is presumed, will prove entertaining and useful to the reader.

"For what can be more interesting to a person in a long voyage, than to be able to tell upon what part of the globe he is, to know how far he has travelled, what distance he has to go, and how he must direct his course to arrive at the place he designs to visit? These important particulars are all determined by knowing the latitude and longitude of the place under consideration. When the discovery of the compass invited the voyager to quit his native shore, and venture himself upon an unknown ocean, that knowledge, which before he deemed of no importance, now became a matter of absolute necessity. Floating in a frail vessel, upon an uncertain abyss, he has consigned himself to the mercy of the winds and waves, and knows not where he is."*

The following instance will prove of what use it is to know the longitude of places at sea. The editor of Lord Anson's voyage, speaking of the island of Julian Fernandez, adds, " The uncertainty we were in of it's position, and our standing in for the main on the 28th of May, in order to secure a sufficient easting, when we were indeed extremely near it, cost us the lives of between 70 and 80 of our men, by our longer continuance at sea; from which fatal accident we might have been exempted, had

* Bonnycastle's Astronomy.
230

we been furnished with such an account of it's situation, as we could fully have depended on."

The latitude of a place the sailor can easily discover; but the longitude is a subject of the utmost difficulty, for the discovery of which many methods have been devised. It is indeed of so great consequence, that the Parliament of Great Britain proposed a reward of 10,000 *l.* if it extended only to 1 degree of a great circle, or 60 geographical miles; 15,000 *l.* if found to 40 such miles; and 20,000 *l.* to the person that can find it within 30 minutes of a great circle, or 30 geographical miles.

As I cannot enter fully into this subject in these essays, it will, I hope, be deemed sufficient, if I give such an account as will enable the reader to form a general idea of the solution of this important problem.

From what has been seen in the preceding pages, it is evident that 15 degrees in longitude answer to one hour in time, and consequently that the longitude of any place would be known, if we knew their difference in time; or in other words, how much sooner the sun, &c. arrives at the meridian of one place, than that of another, The hours and degrees being in this respect commensurate, it is as proper to express the distance of any place in time as in degrees.

Now it is clear, that this difference in time would be easily ascertained by the observation of any instantaneous appearance in the heavens, at two distant places ; for the difference in time at which the same phenomenon is observed, will be the distance of the two places from each other in longitude. On this principle, most of the methods in general use are founded.

Thus if a clock, or watch, was so contrived, as to go uniformly in all seasons, and in all places; such a watch being regulated to London time, would always shew the time of the day at London ; then the time of the day under any other meridian being found, the difference between that time, and the corresponding London time, would give the difference in longitude.

For supposing any person possessed of one of these time-pieces, to set out on a journey from London, if his time piece be accurately adjusted, wherever he is, he will always know the hour at London exactly; and when he has proceeded so far either eastward or westward, that a difference is perceived betwixt the hour shewn by his time-piece, and those of the clocks and watches at the places to which he goes, the distance of those places from London in longitude will be known. But to whatever degree of perfection such movements may be made, yet as every mechanical instrument is liable to be injured by various accidents, other methods are obliged to be used, as the eclipses of the sun and moon, or of Jupiter's satellites. Thus supposing the moment of the beginning of an eclipse was at ten o'clock at night at London, and by accounts from two observers in two other places, it appears that it began with one of them at nine o'clock, and with the other at midnight; it is plain, that the place where it began at nine is one hour, or 15 degrees east in longitude from London; the other place where it began at midnight, is 30 degrees distant in west longitude from London. Eclipses of the sun and moon do not, however, happen often enough to answer the purposes of navigation; and the motion of a ship at sea prevents the observations of those of Jupiter's satellites.

If the place of any celestial body be computed, for example, as in an almanack, for every day or to parts of days, to any given meridian, and the place of this celestial body can be found by observation at sea, the difference of time between the time of observation and the computed time, will be the difference of longitude in time. The moon is found to be the most proper celestial object, and the observations of her appulses to any fixed star is reckoned one of the best methods for resolving this difficult problem.

LENGTH OF THE DEGREES OF LONGITUDE.

Supposing the earth to be a perfect globe, the length of a degree upon the meridian has been estimated to be 69,1 miles; but as the earth is an oblate spheroid, the length of a degree on the equator will be somewhat greater.

Whether the earth be considered as a spheroid or a globe, all the meridians intersect one another at the poles. Therefore, the number of miles in a degree must always decrease as you go north or south from the equator. This is evident by inspection of a globe, where the parallels of latitude are found to be smaller in proportion as they are nearer the pole. Hence it is that a degree of longitude is no where the same, but upon the same parallel; and that a degree of longitude is equal to a degree of latitude only upon the equator.

The following *table* shews how many geographical miles, and decimal parts of a mile, would be contained in a degree of longitude, at each degree of latitude from the equator to the poles, if the earth was a perfect sphere, and the circumference of it's equinoctial line 360 degrees, and each degree 60 geographical miles.

This table enables us to determine the velocity with which places upon the globe revolve eastward; for the velocity is different, according to the distance of the places from the equator, being swiftest as passing through a greater space, and so by degrees slower towards the pole, as passing through a less space in the same time. Now as every part of the earth is moved through the space of it's circumference, or 360 degrees, in 24 hours; the space described in one hour is found by deviding 360 by 24, which gives in the quotient 15 degrees; and so many degrees does every place on the earth move in an hour. The number of miles contained in so many degrees in any latitude, is readily found from the table.

Thus under the equator places revolve at the rate of more than 1000 miles in an hour; at London, at the rate of about 640 miles in an hour.

TABLE.

LAT. Deg.	Miles.	LAT. Deg.	Miles.	LAT. Deg.	Miles.
00	60,00	10	59,08	20	56,38
1	59,99	11	58,89	21	56,01
2	59,96	12	58,68	22	55,63
3	59,92	13	58,46	23	55,23
4	59,86	14	58,22	24	54,81
5	59,77	15	57,95	25	54,38
6	59,67	16	57,67	26	53,93
7	59,56	17	57,37	27	53,46
8	59,42	18	57,06	28	52,97
9	59,26	19	56,73	29	52,47

LAT.		LAT.		LAT.	
Deg.	*Miles.*	*Deg.*	*Miles.*	*Deg.*	*Miles.*
30	51,96	51	37,76	72	18,55
31	51,43	52	36,94	73	17,54
32	50,88	53	36,11	74	16,53
33	50,32	54	35,26	75	15,52
34	49,74	55	34,41	76	14,51
35	49,15	56	33,55	77	13,50
36	48,54	57	32,68	78	12,47
37	47,92	58	31,79	79	11,45
38	47,28	59	30,90	80	10,42
39	46,62	60	30,00	81	9,38
40	45,95	61	29,09	82	8,35
41	45,28	62	28,17	83	7,32
42	44,59	63	27,24	84	6,28
43	43,88	64	26.30	85	5,23
44	43,16	65	25,36	86	4,18
45	42,43	66	24,41	87	3,14
46	41,68	67	23,45	88	2,09
47	40,92	68	22,48	89	1,05
48	40,15	69	21,50	90	0,00
49	39,36	70	20,52		
50	38,57	71	19,54		

Another circumstance which arises from this difference of meridians in time, must detain us a little before we quit this subject. For from this difference it follows, that if a ship sails round the world, always directing her course eastward, she will at her return home find she has gained one whole day of those that stayed at home; that is, if they reckon it May 1, the ship's company will reckon it May 2; if westward, a day less, or April 30.

This circumstance has been taken notice of by navigators. " It was during our stay at Mindanao, (says Capt. Dampier) that we were first made sensible of the change of time in the course of our voyage: for having travelled so far westward, keeping the same course with the sun, we consequently have gained something insensibly in the length of the particular days, but have lost in the tale the bulk or number of the days or hours.

" According to the different longitudes of England and Mindanao, this isle being about 210 degrees west from the Lizard, the difference of time at our arrival at Mindanao ought to have been about fourteen hours; and so much we should have anticipated our reckoning, have gained it by bearing the sun company.

" Now the natural day in every place must be consonant to itself; but going about with, or against the sun's course, will of necessity make a difference in the calculation of the civil day, between any two places. Accordingly, at Mindanao, and other places in the East Indies, we found both natives and Europeans reckoning a day before us. For the Europeans coming eastward, by the Cape of Good Hope, in a course contrary to the sun and us, wherever we met, were a full day before us in their accounts.

"So among the Indian Mahometans, their Friday was Thursday with us; though it was Friday also with those that came eastward from Europe.

"Yet at the Ladrone islands we found the Spaniards of Guam keeping the same computation with ourselves; the reason of which I take to be, that they settled that colony by a course westward from Spain; the Spaniards going first to America and thence to the Ladrone islands."

It is clear, from what has been said in the first part of this article, concerning both latitude and longitude, that if a person travel ever so far directly towards east or west, his latitude would be always the same, though his longitude would be continually changing.

But if he went directly north or south, his longitude would continue the same, but his latitude would be perpetually varying.

If he went obliquely, he would change both his latitude and longitude.

The longitude and latitude of places give only their relative distances on the globe; to discover, therefore, their real distance, we have recourse to the following problem.

PROBLEM X.

Any place being given, to find the distance of that place from another, in a great circle of the earth.

I shall divide this problem into three cases.

Case 1. If the places lie under the same meridian. Bring them up to the meridian, and mark the number of degrees intercepted between them. Multiply the number of degrees thus found by 60, and they will give the number of geographical miles between the two places. But if we would have the number of English miles, the degrees before found must be multiplied by $69\frac{1}{2}$.

Case 2. If the places lie under the equator. Find their difference of longitude in degrees, and multiply, as in the preceding case, by 60 or $69\frac{1}{2}$.

Case 3. If the places lie neither under the same meridian, nor under the equator. Then lay the quadrant of altitude over the two places, and mark the number of degrees intercepted between them. These degrees multiplied as above mentioned, will give the required distance.

PROBLEM XI.

To find the angle of position of places.

The angle of position is that formed between the meridian of one of the places, and a great circle passing through the other place.

Rectify the globe to the latitude and zenith of one of the places, bring that place to the strong brass meridian, set the graduated edge of the quadrant to the other place, and the number of degrees contained between it and the strong brass meridian, is the measure of the angle sought. Thus,

The angle of position between the meridian of Cape Clear, in Ireland, and St. Augustine, in Florida, is about 82 degrees south westerly; but the angle of position between St. Augustine and Cape Clear, is only about 46 degrees north easterly.

Hence it is plain, that the line of position, or azimuth, is not the same from either place to the other, as the romb-line are.

PROBLEM XII.

To find the bearing of one place from another.

The bearing of one sea-port from another is determined by a kind of spiral, called a romb-line, passing from one to the other, so as to make equal angles with all the meridians it passes by; therefore, if both places are situated on the same parallel of latitude, their bearing is either east or west from each other; if they are upon the same meridian, they bear north and south from one another; if they lie upon a romb-line, their bearing is the same with it; if they do not, observe to which romb-line the two places are nearest parallel, and that will shew the bearing sought.

Example. Thus the bearing of the Lizard point from the island of Bermudas is nearly E. N. E.; and that of Bermudas from the Lizard is W. S. W. both nearly upon the same romb-line, but in contrary directions.

OF THE TWILIGHT.

That light which we have from the sun before it rises, and after it sets, is called the *twilight*.

The morning twilight, or day break, commences when the sun comes within eighteen degrees of the horizon, and continues till sun-rising. The evening twilight begins at sun-setting, and continues till it is eighteen degrees below the horizon.

To illustrate the causes of the various length of twilight in different places, a wire circle is fixed eighteen degrees below the surface of the broad paper circle; so that all those places which are above the wire circle will have twilight, but it will be dark to all those places below it.

I have already observed, that it is owing to the atmosphere that we are favoured with the light of the sun before he is above, and after he is below, our horizon. Hence, though after sun-setting we receive no direct light from the sun, yet we enjoy his reflected light for some time; so that the darkness of the night does not come on suddenly, but by degrees.

In a right position of the sphere the twilights are quickly over, because the sun rises and sets nearly in a perpendicular; but in an oblique sphere they last longer, the sun rising and setting obliquely. The greater the latitude of the place, the longer is the duration of the twilight; so that all those who are in 49 degrees of latitude have in the summer, near the solstice, their atmosphere enlightened the whole night, the twilight lasting till sun-rising.

In a parallel sphere, the twilight lasts for several months; so that the inhabitants of this position have either direct or reflex light of the sun nearly all the year, as will plainly appear by the globe.

OF THE DIURNAL MOTION OF THE EARTH, AND THE PROBLEMS DEPENDING ON THAT MOTION.

As the daily motion of the earth about it's axis, and the phenomena dependent on it, are some of the most essential points which a beginner ought to have in view, we shall now endeavour to explain them by the globes; and here I think the advantage of globes mounted in my father's manner, over those generally used, will be very evident.

I have already observed, that in globes mounted in our manner, the motion of the terrestrial globe about it's axis represents the diurnal motion of the earth, and that the horary index will point out upon the equator the 24 hours of one diurnal rotation, or any part of that time.

I shall now consider *the broad paper circle as the plane which distinguishes light from darkness;* that is, the enlightened half of the earth's surface, from that which is not enlightened.

For when the sun shines upon a globe, he shines only upon one half of it; that is, one half of the globe's surface is enlightened by him, the other not.

That the enlightened half may be that half

which is above the broad paper circle, we must imagine the sun to be in our *zenith*.

Or let a sun be painted on the ceiling over the terrestrial globe, the diameter of the picture equal to the diameter of the globe.

Then all those places that are above the broad paper circle will be in the sun's light; that is, it will be *day* in all those places.

And all places that are below this circle, will be out of the sun's light; that is, in all those places it will be *night*.

When any place on the earth's surface comes to the edge of the broad paper circle, passing out of the shade into the light, the sun will appear *rising* at that place.

And when a place is at the edge of the broad paper circle, going out of the light into the shade, the sun will appear at that place to be *setting*.

When we view the globe in this position, we at once see the situation of all places in the illuminated hemisphere, whose inhabitants enjoy the light of the day. One edge of the broad paper circle shews at what place the sun appears rising at the *same* time; and the opposite edge shews at what places the sun is setting at the same time.

The horary index shews how long a place is moving from one edge to the other; that is, how long the day or night is at that place;

and, consequently, when the globe is thus situated, you readily discover the time of the sun's rising and setting on any given day, in any given place.

TO RECTIFY THE TERRESTRIAL GLOBE.

To rectify the terrestrial globe, is to place it in the same position in which our earth stands to the sun, at all or at any given times.

That half of the earth's surface which is enlightened by the sun is not always the same; it differs according as the sun's declination differs.

To rectify, then, the terrestrial globe, is to bring it into such a position, as that the enlightened half of the earth's surface may be all above the broad paper circle.

On the back side of the strong brass meridian, and on each side of the north pole, the months and days of the month are graduated in two concentric spaces, agreeable to the declination of the sun.

Bring the day of the month that is graduated on the back side of the strong brass meridian, to coincide with the broad paper circle, and the globe is rectified.

Thus set the first of May to coincide with the broad paper circle, and that half of the earth's surface which is enlightened at any time upon that day, will be all at once above the said circle.

If the horary index be set to XII, when any particular place is brought under the strong brass meridian, it will shew the precise time of sun-rising and sun-setting at that place, according as that place is brought to the eastern or western edge of the broad paper circle.

It will also shew how long any place is in moving from the east to the west side of the illuminated disk, and thence the length of the day and night.

It will also point out the length of the twilight, by shewing the time in which the place is passing from the twilight circle to the edge of the broad paper circle on the western side; or from the edge of this circle on the eastern side, to the twilight wire, and thus determine the length of the whole artificial day.

N. B. The twilight wire is placed at 18 degrees from the broad paper circle.

I shall now proceed to exemplify upon the globes these particulars, at three different seasons of the year, viz. the summer solstice, the winter solstice, and the time or times of the equinoxes.

PROBLEM XIII.

To place the globe in the same situation, with respect to the sun, as our earth is in at the time of the SUMMER SOLSTICE.

Rectify the globe to the extremity of the divisions for the month of June, or $23\frac{1}{2}$ degrees north declination; that is, bring these divisions on the strong brass meridian to coincide with the plane of the broad paper circle.

Then that part of the earth's surface, which is within the northern polar circle, will be above the broad paper circle, and will be in the light, and the inhabitants thereof will have no night.

But all that space which is contained within the southern polar circle, will continue in the shade; that is, it will there be continual night.

In this position of the globe, the pupil will observe how much the diurnal arches of the parallels of latitude decrease, as they are more and more distant from the elevated pole.

If any place be brought under the strong brass meridian, and the horary index is set to that XII which is most elevated, and the place be afterwards brought to the western side of the broad paper circle, the hour index will shew the time of sun-rising; and when the place is moved to the eastern edge, the index points to the time of sun-setting.

The length of the day is obtained by the time shewn by the horary index, while the globe moves from the west to the east side of the broad paper circle.

Thus it will be found, that at London the sun rises about 15 minutes before IV in the morning, and sets about 15 minutes after VIII at night.

At the following places it will be nearly at the times expressed in the table.

	☉ Rising.		☉ Setting.		Length of day.		Twilight.	
	h.	m	h.	m.	h.	m.	h.	m.
Cape Horn	8	44	3	16	6	32	2	35
Cape of Good Hope	7	9	4	51	9	42	1	43
Rio de Janeiro, in Brazil	6	42	5	19	10	38	1	23
Island of St. Thomas's near the equator.	6		6		12		1	20
Cape Lucas, California	5	12	6	48	13	36	1	35

We also see, that at the same time the sun is rising at London, it is rising at the isles of Sicily and Madagascar.

And, that at the same time when the sun sets at London it is setting at the island of Madeira, and at Cape Horn.

And when the sun is setting at the island of Borneo, in the East Indies, it is rising at Florida, in America. And many other similar circumstances relative to other places, are seen as it were by inspection.

PROBLEM XIV.

To explain the situation of the earth, with respect to the sun, at the time of the WINTER SOLSTICE.

Rectify the globe to the extremity of the divisions for the month of December, or to $23\frac{1}{2}$ degrees south declination.

When it will be apparent that the whole space within the southern polar circle is in the sun's light, and enjoys continual day; whilst that of the northern polar circle is in the shade, and has continual night.

If the globe be turned round, as before, the horary index will shew, that at the several places before-mentioned their days will be respectively equal to what their nights were at the time of the summer solstice.

It will appear farther, that it is now sun-setting at the same time in those places in which it was sun-rising at the same time at the summer solstice; and, on the contrary, sun-rising at the time it then appeared to set.

PROBLEM. XV.

To place the globe in the situation of the earth, at the times of the EQUINOX.

The sun has no declination at the times of the equinox, consequently there must be no elevation of the pole.

Bring the day of the month when the sun enters the first point of Aries, or day of the month when the sun enters the first point of Libra, to the plane of the broad paper circle; then the two poles of the globe will be in that plane also, and the globe will be in the position which is called a *right sphere*.

For it is a right sphere when the two poles are in the plane of the broad paper circle, because then all those circles which are parallel to the equator will be at right angles to that plane.

If the globe be now turned from west to east, it will plainly appear, that all places upon it's surface are twelve hours above the broad paper circle, and twelve hours below it; that is, the days are twelve hours long all over the earth, and the nights are equal to the days, whence these times are called the times of equinox.

Two of these occur in every year; the first

is the autumnal, the second the vernal equinox.

At these seasons the sun appears to rise at the same time to all places that are on the same meridian. The sun sets also at the same time in all those places.

Thus if London and Mundford, on the gold coast, be brought to the strong brass meridian, the graduated side of which is in this case the horary index, and they be afterwards carried to the western edge of the broad paper circle, the index will shew that the sun rises at VI at both places; when they are carried to the eastern edge, the index points to VI for the time of sunsetting.

N. B. If London be not the given place, the hour index is to be set to the most elevated XII, while the place is under the graduated edge of the strong brass meridian.

The following circumstances, which usually attend the four cardinal divisions of the year, cannot be better introduced than at this place. At the time of the equinoxes, when the sun passes from one hemisphere into the other, there is almost constantly some disturbance in the weather; the winds are then generally higher: at the vernal equinox they are for the most part easterly, cold, dry, and searching. The solstitial point of the summer is often distinguished by violent rains, and that we call a midsummer flood. The winter being less rainy than the summer, nothing particular happens at the winter solstice, but that the frosts commonly set in more severely, with some quantity of snow upon the ground.

OF THE ARTIFICIAL OR TERRESTRIAL HORIZON.

The brass circle, which may be slipped from pole to pole on the moveable meridian, has been already described. The circumference of it is divided into eight parts, to which are affixed the initial letters of the mariner's compass.

When the center of it is set to any particular place, the situation of any other place is seen, with respect to that place; that is, whether they be east, west, north, or south of it.

It will therefore represent the horizon of that place.

We shall here use this artificial horizon, to shew why the sun, although he be always in one and the same place, appears to the inhabitants of the earth at different altitudes, and in different azimuths.

PROBLEM XVI.

To exemplify the sun's altitude, as observed with an artificial horizon.

The altitude of the sun is greater or less, according as the line which goes from us to the sun is nearer to, or farther off from our horizon.

Let the moveable circle be applied to any place, as London, then will the horizon of London be thereby represented.

The sun is supposed, as before, to be in the zenith, that is, directly over the terrestrial globe.

If then from London a line go vertically upwards, the sun will be seen at London in that line.

At sun-rising, when London is brought to the west edge of the broad paper circle, the supposed line will be parallel to the artificial horizon, and the sun will then be seen in the horizon.

As the globe is gradually turned from the west towards the east, the horizon will recede from that line which goes from London vertically upwards; so that the line in which the sun is seen gets further and further from the horizon; that is, the sun's altitude increases gradually.

When the horizon, and the line which goes from London vertically upwards, are arrived at the strong brass meridian, the sun is then at his greatest or meridian altitude for that day, and the line and horizon are at the largest angle they can make with each other.

After this, the motion of the globe being continued, the angle between the artificial horizon, and the line which goes from London vertically upwards, continually decreases, until London arrives at the eastern edge of the broad paper circle; it's horizon then becomes vertical again, and parallel to the line which goes vertically upwards. The sun will again appear in the horizon, and will set.

PROBLEM XVII.

Of the sun's meridian altitude, at the three different seasons.

Rectify the globe to the time of the winter solstice, by problem xiv, and place the center of the visible horizon on London.

When London is at the graduated edge of the strong brass meridian, the line which goes vertically upwards makes an angle of about 15 degrees; this is the sun's meridian altitude at that season, to the inhabitants of London.

If the globe be rectified to the times of equinox, by problem xv, the horizon will be

OF THE GLOBES. 69

farther separated from the line which goes vertically upwards, and makes a greater angle therewith, it being about 38½ degrees; this is the sun's meridian altitude, at the time of equinox at London.

Again, rectify to the summer solstice by problem xiii, and you will find the artificial horizon recede farther from the line which goes from London vertically upwards, and the angle it then makes is about 62 degrees, which shews the sun's meridian altitude at the time of the summer solstice.

Hence flows also the following arithmetical problem.

PROBLEM XVIII.

To find the sun's meridian altitude universally.

Add the sun's declination to the elevation of the equator, if the latitude of the place, and the declination of the sun, are both on the same side.

If on contrary sides, subtract the declination from the elevation of the equator, and you obtain the sun's meridian altitude.

Thus the elevation of the equator at
London is - - - 38° 28
The sun's declination on the 20th of
May - - - - 20 8
Their sum, the sun's meridian altitude
that day - - - - 58 36

70   DESCRIPTION AND USE

Again, to the elevation of the equator
at London - - - 38° 28
Add the sun's greatest declination at
the time of the summer solstice 23 29

The sum is the sun's greatest meridian
altitude at London - - 61 57

PROBLEM XIX.

Of the sun's azimuths, as compared with the artificial horizon.

The artificial horizon serves also to determine the sun's azimuths.

An *azimuth* of the sun is denominated from that point of the horizon, to which the sun, or a line going to the sun, is nearest.

Thus if the sun, or a line going to the sun, be nearest the south-east point of the horizon, which point is 45 degrees distant from the meridian, the sun's azimuth is an azimuth of 45 degrees, and the sun will appear in the south-east.

Imagine the sun, as we have done before, to be placed directly over the globe.

In which case, a line going to the sun from any place on the surface of the globe, will have a vertical direction, and will go from that place vertically upwards.

OF THE GLOBES. 71

If then we apply the artificial horizon to any place, the point of this horizon to which a vertical line is nearest, shews the sun's azimuth at that time.

It is observable, that the point of the horizon to which such a vertical line is nearest, will be at all times that point which is most elevated.

To exemplify this, let the globe be in the position of a right sphere, and let the artificial horizon be applied to London.

When London is at the western edge of the broad paper circle, which situation represents the time when the sun appears to rise, the eastern point of the artificial horizon being then most elevated, shews that the sun at his rising is due east.

Turn the globe, till London comes to the eastern edge of the broad paper circle, then the western point of the artificial horizon will be most elevated, shewing that the sun sets due west.

Now place the globe in the position of an oblique sphere; and if London be brought to the eastern or western side of the broad paper circle, the vertical line will depart more or less from the east and west points, in which case the sun is said to have more or less *amplitude*.

If the departure be northward, it is called

northern amplitude; if southward, it is called southern amplitude.

In whatever position the globe be placed,* when London comes to the strong brass meridian, the most elevated part of the artificial horizon will be the south point of it.

Which shews that at noon the sun will always, and in all seasons, appear in the south.

OF THE ANCIENT DIVISIONS OF THE EARTH INTO ZONES AND CLIMATES.

Climates was a term used by the ancient astronomers to express a division of the earth, which, before the marking down the latitudes of countries into degrees and minutes was in use, served them for dividing the earth into certain portions in the same direction, so as to speak of any particular place with some degree of certainty, though not with due precision.

It was natural for the earliest observers to remark, for one of the first things, the diversity that there was in the sun's rising and setting: it was by this they regulated what they called climates; which are a tract on the surface of the earth, of various breadths, being regulated by the different lengths of time be-

* The globe is not supposed in this case, or under this view of things, ever to be elevated above the limits of the sun's declination.

OF THE GLOBES. 73

tween the rising and setting of the sun in the longest day, in different places.

From the equator to the latitude 66½ north and south, a climate is constituted by the difference of half an hour in the length of the longest day, and this is sufficient for understanding the ancients. Between the polar circle and the pole, the length of the longest day, in one parallel, exceeds the length of the longest in the next by a month; but of these the ancients knew nothing.

CLIMATES BETWEEN THE EQUATOR AND POLAR CIRCLES.

Climates	Hours.	Latitude. D. M.	Breadth. D. M.	Climates	Hours.	Latitude. D. M.	Breadth. D. M.
1	12½	8 25	8 25	13	18½	59 58	1 29
2	13	16 25	8 00	14	19	61 18	1 20
3	13½	23 50	7 25	15	19½	62 25	1 07
4	14	30 25	6 30	16	20	63 22	0 57
5	14½	36 28	6 08	17	20½	64 06	0 44
6	15	41 22	4 54	18	21	64 49	0 43
7	15½	45 29	4 07	19	21½	65 21	0 32
8	16	49 01	3 32	20	22	65 47	0 22
9	16½	52 00	2 57	21	22½	66 06	0 19
10	17	54 27	2 29	22	23	66 20	0 14
11	17½	56 37	2 10	23	23½	66 28	0 08
12	18	58 29	1 52	24	24	66 31	0 03

Therefore, to discover in what climate a place is, whose latitude does not exceed 66½ degrees, find the length of the longest day in that place, and subtracting 12 hours from that length, the number of half hours in the remainder will specify the climate.

PROBLEM XX.

To find the limits of the climates.

Elevate the north pole to 23° 28′, the sun's declination on the longest day; and turn the globe easterly till the intersection of the meridian with the equator that passes through Libra comes to the horizon, and the hour of VI will then be under the meridian, which in this problem is the hour index, because the sun sets this day at places on the equator as it does every day at VI o'clock. Now turn the globe easterly till the time under the meridian is 15 min. past VI. and you find that 8° 34′ of that graduated meridian is cut by the horizon; this is the beginning of the second climate; and the limits of all the climates may be determined, by bringing successively the time equal to half the length of the longest day under the meridian, and observing the degree of the graduated meridian cut by the horizon.

ZONES.

Zones is another division of the earth's surface used by the ancients: that part which the sun passes over in a year, comprehending 23¼ degrees on each side the equator, was called by the ancients the torrid zone. The two frigid zones are contained between the polar circles. Between the torrid and the two frigid zones are contained the two temperate ones, each being about 43 degrees broad.

The latitude of a place being the mark of it's position with respect to the sun, may be considered as a general index to the temperature of the climate: it is, however, liable to very great exceptions; but to deny it absolutely would be to deny that the sun is the source of light and heat below.

Nothing can be more hideous or mournful than the pictures which travellers present us of the polar regions. The seas, surrounding inhospitable coasts, are covered with islands of ice, that have been increasing for many centuries: some of these islands are immersed six hundred feet under the surface of the sea, and yet often rear up also their icy heads more than one hundred feet above it's level, and are three or four miles in circumference. The following account will give some idea of the scenery produced by arctic weather. At Smearingborough-Harbour, within fifteen degrees of the pole, the country is full of mountains, precipices, and rocks; these are covered with ice and snow. In the vallies are hills of ice, which seem daily to accumulate. These hills assume many strange and fantastic appearances; some looking like churches or castles, ruins, ships in full sail, whales, monsters, and all the various forms that fill the universe. There are seven of these ice-hills, which are the highest in the country. When the air is clear, and the light shines full upon them, the prospect is inconceivably brilliant; the sun is reflected from them as from glass; sometimes they appear of a bright hue, like sapphire; sometimes variegated with all the glories of the prismatic colours, exceeding, in the magnitude of lustre, and beauty of colour, the richest gems in the world, disposed in shapes wonderful to behold, dazzling the eye with the brilliancy of it's splendor. At Spitsbergen, within ten degrees of the pole, the earth is locked up in ice till the middle of May; in the beginning of July the plants are in flower, and perfect their seeds in a month's time: for though the sun is much more oblique in the higher latitudes than with us, his long continuance above the horizon is attended with an accumulation of heat exceeding that of many places under the torrid zone; and there is reason to suppose, that the rays of the sun, at any given altitude, produce greater degrees of heat in the condensed air of the polar regions, than in the thinner air of this climate.

Yet, if we look for heat, and the remarkable effects of it, we must go to the countries near the equator, where we shall find a scenery totally different from that of the frigid zone. Here all things are upon a larger scale than in the temperate climates; their days are burning hot; in some parts their nights are piercing cold; their rains lasting and impetuous, like torrents; their dews excessive; their thunder and lightning more frequent, terrible, and dangerous; the heat burns up the lighter soil, and forms it into a sandy desert, while it quickens all the moister tracts with incredible vegetation.

The ancients supposed that the frigid zone was uninhabitable from cold, and the torrid from the intolerable heat of the sun; we now, however, know that both are inhabited. The sentiments of the ancients, therefore, in this respect, are a proof how inadequate the faculties of the human mind are to discussions of this nature, when unassisted by facts.

OF THE ANCIENT DISTINCTION OF PLACES, BY THE DIVERSITY OF SHADOWS OF UPRIGHT BODIES AT NOON.

When the sun at noon is in the zenith of any place, the inhabitants of that place were by the ancients called *ascii*, that is, without shadow; for the shadow of a man standing upright, when the sun is directly over his head, is not extended beyond that part of the earth which is directly under his body, and therefore will not be visible.

As the shadow of every opake body is extended from the sun, it follows, that when the sun at noon is southward from the zenith of any place, the shadow of an inhabitant of that place, and indeed of any other opake body, is extended towards the north.

But when the sun is northward from the zenith of any place, the shadow falls towards the south.

Those are called *amphiscii*, that have both kinds of meridian shadows.

Those, whose meridian shadows are always projected one way, are termed *heteroscii*.

PROBLEM XXI.

To illustrate the distinction of ascii, amphiscii, heteroscii, and periscii, by the globe.

Rectify the globe to the summer solstice, and move the artificial horizon to the equator, the north point will be the most elevated at noon.

Which shews, that to those inhabitants who live at the equator, the sun will at this season appear to the north at noon, and their shadow will therefore be projected southwards.

But if you rectify the globe to the winter solstice, the south point being then the uppermost point at noon, the same persons will at noon have the sun on the south side of them, and will project their shadows northwards.

Thus they are amphiscii, projecting their shade both ways; which is the case of all the inhabitants within the tropics.

The artificial horizon remaining as before, rectify the globe to the times of the equinox, and you will find that when this horizon is under the strong brass meridian, a line going vertically upwards will be perpendicular to it, and consequently the sun will be directly over the heads of the inhabitants, and they will be ascii, having no noon shade; their shadow is in the morning projected directly westward, in the evening directly eastward.

The same thing will also happen to all the inhabitants who live between the tropics of Cancer and Capricorn; so that they are not only ascii, but amphiscii also.

Those who live without the tropics are heteroscii; those in north latitude have the noon shade always directed to the north, while those in south latitude have it always projected to the south.

The inhabitants of the polar circles are

called *periscii;* because, as the sun goes round them continually, their shade goes round them likewise.

OF ANCIENT DISTINCTIONS FROM SITUATION.

These terms being often mentioned by ancient geographical writers to express the different situation of parts of the globe, by the relation which the several inhabitants bore to one another, it will be necessary to take some notice of them.

The *antœci* are two nations which are in or near the same meridian; the one in north, the other in south latitude.

They have therefore the same longitude, but not the same latitude; opposite seasons of the year, but the same hour of the day; the days of the one are equal to the nights of the other, and, *vice versa,* when the days of the one are at the longest, they are shortest at the other.

When they look towards each other, the sun seems to rise on the right hand of the one, but on the left of the other. They have different poles elevated; and the stars that never set to the one, are never seen by the other.

Periœci are also two opposite nations, situated on the same parallel of latitude.

They have therefore the same latitude, but differ 180 degrees in longitude; the same seasons of the year, but opposite hours of the day; for when it is twelve at night to the one, it is twelve at noon with the other. On the equinoctial days, the sun is rising to one, when it is setting to the other.

Antipodes are two nations diametrically, opposite, which have opposite seasons and latitude, opposite hours and longitude.

The sun and stars rise to the one, when they set to the other, and that during the whole year, for they have the same horizon.

The day of the one is the night of the other; and when the day is longest with the one, the other has it's shortest day.

They have the contrary seasons at the same time; different poles, but equally elevated; and those stars that are always above the horizon of one, are always under the horizon of the other.

PROBLEM XXII.

To find the Antœci, the Periœci, and the Antipodes of any place.

Bring the given place to the strong brass meridian, then in the opposite hemisphere, and under the same degree of latitude with the given place, you will find the antœci.

The given place remaining under the meridian, set the horary index to XII; then turn the globe, till the other XII is under the index, then will you find the periœci under the same degree of latitude with the given place.

Thus the inhabitants of the south part of Chili are antœci to the people of New England, whose Periœci are those Tartars who dwell on the north borders of China, which Tartars have the said inhabitants of Chili for their antipodes.

This will become evident, by placing the globe in the position of a right sphere, and bringing those nations to the edge of the broad paper circle.

PROBLEM XXIII.

The day of the month being given, to find all those places on the globe, over whose zenith the sun will pass on that day.

Rectify the terrestrial globe, by bringing the given day of the month on the back side of the strong brass meridian, to coincide with the plane of the broad paper circle; observe the number of degrees of the brass meridian, which corresponds to the given day of the month.

This number of degrees, counted from the equator on the strong brass meridian, towards the elevated pole, is the point over which the sun is vertical; and all those places, which pass under this point, have the sun directly vertical on the given day.

Example. Bring the 11th of May to coincide with the plane of the broad paper circle, and the said plane will cut eighteen degrees for the elevation of the pole, which is equal to the sun's declination for that day, which being counted on the strong brass meridian towards the elevated pole, is the point over which the sun will be vertical; and all places that are under this degree, will have the sun on their zenith on the 11th of May.

Hence, when the sun's declination is equal to the latitude of any place in the torrid zone, the sun will be vertical to those inhabitants that day; which furnishes us with another method of solving this problem.

OF PROBLEMS PECULIAR TO THE SUN.

PROBLEM XXIV.

To find the sun's place on the broad paper circle.

Consider whether the year in which you seek the sun's place is bissextile, or whether it is the first, second, or third year after.

If it be the first year after bissextile, those divisions to which the numbers for the days of the months are affixed, are the divisions which are to be taken for the respective days of each month of that year at noon; opposite to which, in the circle of twelve signs, is the sun's place.

If it be the second year after bissextile, the first quarter of a day backwards, or towards the left hand, is the day of the month for that year, against which, as before, is the sun's place.

If it be the third year after bissextile, then three quarters of a day backwards is the day of the month for that year, opposite to which is the sun's place.

If the year in which you seek the sun's place be bissextile, then three quarters of a day backwards is the day of the month from the 1st of January to the 28th of February inclusive. The intercalary, or 29th day, is three-fourths of a day to the left hand from the 1st of March, and the 1st of March itself one quarter of a day forward, from the division marked 1; and so for every day in the remaining part of the leap year; and opposite to these divisions is the sun's place.

In this manner the intercalary day is very well introduced every fourth year into the calendar, and the sun's place very nearly obtained, according to the Julian reckoning.

Thus,

A. D		Sun's place.	Apr. 25.
1788	Bissextile - - -	8	5° 35
1789	First year after - -	8	5 21
1790	Second - - -	8	5 6
1791	Third - - - -	8	4 55

Upon my father's globes there are twenty-three parallels, drawn at the distance of one degree from each other on both sides the equator, which, with two other parallels at $23\frac{1}{2}$ degrees distance, include the ecliptic circle.

The two outermost circles are called the tropics; that on the north side the equator is called the tropic of Cancer, that which is on the south side, the tropic of Capricorn.

Now as the ecliptic is inclined to the equator, in an angle of $23\frac{1}{2}$ degrees, and is included between the tropics, every parallel between these must cross the ecliptic in two points, which two points shew the sun's place when he is vertical to the inhabitants of that parallel; and the days of the month upon the broad paper circle answering to those points of the ecliptic, are the days on which the sun passes directly over their heads at noon, and which are sometimes called their two midsummer days.

It is usual to call the sun's diurnal paths parallels to the equator, which are therefore aptly represented by the above-mentioned parallel circle; though his path is properly a spiral line, which he is continually describing all the year appearing to move daily about a degree in the ecliptic.

PROBLEM XXV.

To find the sun's declination, and thence the parallel of latitude corresponding thereto.

Find the sun's place for the given day in the broad paper circle, by the preceeding problem, and seek that place in the ecliptic line upon the globe; this will shew the parallel of the sun's declination among the above-mentioned dotted lines, which is also the corresponding parallel of latitude; therefore all those places, through which this parallel passes, have the sun in their zenith at noon on the given day.

Thus on the 23d of May the sun's declination will be about 20 deg. 10 min.; and upon the 23d of August it will be 11 deg. 13 min. What has been said in the first part of this problem, will lead the reader to the solution of the following.

PROBLEM XXVI.

To find the two days on which the sun is in the zenith of any given place that is situated between the two tropics.

That parallel of declination, which passes through the given place, will cut the ecliptic line upon the globe in two points, which denote the sun's place, against which, on the broad paper circle, are the days and months required. Thus the sun is vertical at Barbadoes April 24, and August 18.

PROBLEM XXVII.

The day and hour at any place in the torrid zone being given, to find where the sun is vertical at that time.

Rectify the globe to the day of the month, and you have the sun's declination; bring the given place to the meridian, and set the hour index to XII; turn the globe till the index points to the given hour on the equator; then will the place be under the degree of the declination previously found.

Let the given place be London, and time the 11th day of May, at 4 min. past V in the afternoon; bring the 11th of May to coincide with the broad paper circle, and opposite to it you will find 18 degrees of north declination; as London is the given place, you have only to turn the globe till 4 min. past V westward of it is on the meridian, when you will find Port-Royal, in Jamaica, under the 18th degree of the meridian, which is the place where the sun is vertical at that time.

PROBLEM XXVIII.

The time of the day at any one place being given, to find all those places where at the same instant the sun is rising, setting, and on the meridian, and where he is vertical; likewise those places where it is midnight, twilight, and dark night; as well as those places in which the twilight is beginning and ending; and also to find the sun's altitude at any hour in the illuminated, and his depression in the obscure, hemisphere.

Rectify the globe to the day of the month, on the back side of the strong brass meridian, and the sun's declination for that day; bring the given place to the strong brass meridian, and set the horary index to XII upon the equator; turn the globe from west to east, until the horary index points to the given time. Then

All those places, which lie in the plane of the western side of the broad paper circle, see the sun rising, and at the same time those on the eastern side of it see him setting.

It is noon to all the inhabitants of those places under the upper half of the graduated side of the strong brass meridian, whilst at the same time those under the lower half have mid-night.

All those places which are between the upper surface of the broad paper circle, and the wire circle under it, are in the twilight, which begins to all those places on the western side that are immediately under the wire circle; it ends at all those which are in the plane of the paper circle.

The contrary happens on the eastern side; the twilight is just beginning to those places in which the sun is setting, and it's end is at the place just under the wire circle.

And those places which are under the twilight wire circle have dark night, unless the moon is favourable to them.

All places in the illuminated hemisphere have the sun's latitude equal to their distance from the edge of the enlightened disk, which is known by fixing the quadrant of altitude to the zenith, and laying it's graduated edge over any particular place.

The sun's depression is obtained in the same manner, by fixing the center of the quadrant at the nadir.

PROBLEM XXIX.

To find all those places within the polar circles on which the sun begins to shine, the time he shines constantly, when he begins to disappear, the length of his absence, as well as the first and last day of his appearance to those inhabitants; the day of the month, or latitude of the place being given.

Bring the given day of the month on the back side of the strong brass meridian to the plane of the broad paper circle; the sun is just then beginning to shine on all those places which are in the parallel that just touches the edge of the broad paper circle, and will for several days seem to skim all around, and but a little above their horizon, just as it appears to us at it's setting; but with this observable difference, that whereas our setting sun appears in one part of the horizon only, by them it is seen in every part thereof; from west to south, thence east to north, and so to west again.

Or if the latitude be given, elevate the globe to that latitude, and on the back of the strong brass meridian, opposite to the latitude, you obtain the day of the month; then all the other requisites are answered as above.

As the two concentric spaces, which contain the days of the month on the back side of the strong brass meridian, are graduated to shew the opposite days of the year, at 180 degrees distance; when the given day is brought to coincide with the broad paper circle, it shews when the sun begins to shine on that parallel, which is the first day of it's appearance above the horizon of that parallel.

And the plane of the broad paper circle cuts the day of the month on the opposite concentric space, when the sun begins to disappear to those inhabitants.

The length of the longest day is obtained by reckoning the number of days between the two opposite days found as above, and their difference from 365 gives the length of the longest night.

PROBLEM XXX.

To make use of the globe as a TELLURIAN, *or that kind of orrery which is chiefly intended to illustrate the phenomena that arise from the annual and diurnal motions of the earth.*

Describe a circle with chalk upon the floor, as large as the room will admit of, so that the globe may be moved round upon it; divide this circle into twelve parts, and mark them with the characters of the twelve signs, as they are engraved upon the broad paper circle; placing ♋ at the north, ♑ at the south, ♈ in the east, and ♎ in the west: the mariner's compass under the globe will direct the situation of these points, if the variation of the magnetic needle be attended to.

Note, At London the variation is between 23 and 24 degrees from the north-westward.

Elevate the north pole of the globe, so that $66\frac{1}{2}$ degrees on the strong brass meridian may coincide with the surface of the broad paper circle, and this circle will then represent the plane of the ecliptic, or a plane coinciding with the earth's orbit.

Set a small table, or a stool, over the center of the chalked circle, to represent the sun, and place the terrestrial globe upon it's circumference over the point marked ♑, with the north pole facing the imaginary sun, and the north end of the needle pointing to the variation; and the globe will be in the position of the earth with respect to the sun at the time of the summer solstice, about the 21st of June; and the earth's axis, by this rectification of the globe, is inclined to the plane of the large chalked circle, as well as to the plane of the broad paper circle, in an angle of $66\frac{1}{2}$ degrees; a line, or string, passing from the center of the imaginary sun to that of the globe, will represent a central solar ray connecting the centers of the earth and sun: this ray will fall upon the first point of Cancer, and describe that circle, shewing it to be the sun's place upon the terrestrial ecliptic, which is the same as if the sun's place, by extending the string, was referred to the opposite side of the chalked circle, here representing the earth's path in the heavens.

If we conceive a plane to pass through the center of the globe and the sun's center, it will also pass through the points of Cancer and Capricorn, in the terrestrial and celestial ecliptic; the central solar ray, in this position of the earth, is also in that plane: this can never happen but at the times of the solstice.

If another plane be conceived to pass through the center of the globe at right angles to the center solar ray, it will divide the globe into two hemispheres; that next the center of the chalked circle will represent the earth's illuminated disk, the contrary side of the same plane will at the same time shew the obscure hemisphere.

The reader may realize this second plane by cutting away a semicircle from a sheet of card paste board, with a radius of about $1\frac{1}{2}$ tenth of an inch greater than that of the globe itself.*

If this plane be applied to $66\frac{1}{2}$ degrees upon the strong brass meridian, it will be in the pole of the ecliptic; and in every situation of

* Or he may have a plane made of wood for this purpose.

the globe round the circumference of the chalked circle, it will afford a lively and lasting idea of the various phenomena arising from the parallelism of the earth's axis, and in particular the daily change of the sun's declination, and the parallels thereby described.

Let the globe be removed from ♑ to ♒, and the needle pointing to the variation as before, will preserve the parallelism of the earth's axis; then it will be plain that the string, or central solar ray, will fall upon the first point of Leo, six signs distant from, but opposite to the sign ♒, upon which the globe stands; the central solar ray will now describe the 20th parallel of north declination, which will be about the 23d of July.

If the globe be moved in this manner from point to point round the circumference of the chalked circle, and care be taken at every removal that the north end of the magnetic needle, when settled, points to the degree of variation, the north pole of the globe will be observed to recede from the line connecting the centers of the earth and sun, until the globe is placed upon the point Cancer; after which, it will at every removal tend more and more towards the said line, till it comes to Capricorn again.

PROBLEM. XXXI.

To rectify either globe to the latitude and horizon of any place.

If the place be in north latitude, raise the north pole; if in south latitude, raise the south pole, until the degree of the given latitude, reckoned on the strong brass meridian under the elevated pole, cuts the plane of the broad paper circle; then this circle will represent the horizon of that place, while the place remains in the zenith, but no longer. This rectification is therefore unnatural, though it is the mode adopted in using the globes when mounted in the old manner.

PROBLEM XXXII.

To rectify for the sun's place.

After the former rectification, bring the degrees of the sun's place in the ecliptic line upon the globe to the strong brass meridian, and set the horary index to that XIIth hour upon the equator which is most elevated.

Or if the sun's place is to be retained, to answer various conclusions, bring the graduated edge of the moveable meridian to the degree of the sun's place in the ecliptic, and slide the wire which crosses the center of the artificial horizon thereto; then bring it's center, which is in the intersection of the aforesaid wire, and graduated edge of the moveable meridian, under the strong brass meridian as before, and set the horary index to that XII on the equator which is most elevated.

PROBLEM XXXIII.

To rectify for the zenith of any place.

After the first rectification, screw the nut of the quadrant of altitude so many degrees from the equator, reckoned on the strong brass meridian towards the elevated pole, as that pole is raised above the plane of the broad paper circle, and that point will represent the zenith of the place.

Note, The zenith and nadir are the poles of the horizon, the former being a point directly over our heads, and the latter, one directly under our feet.

If, when the globe is in this state, we look on the opposite side, the plane of the horizon will cut the strong brass meridian at the complement of the latitude, which is also the elevation of the equator above the horizon.

OF THE SOLUTION OF PROBLEMS, BY EXPOSING THE GLOBES TO THE SUN'S RAYS.

In the year 1679, *J. Moxon* published a treatise on what he called " *The English Globe;* being (says he) a stabil and immobil one, performing what the ordinary globes do, and much more; invented and described by the Right Hon. the *Earle of Castlemaine.*" This globe was designed to perform, by being merely exposed to the sun's rays, all those problems which in the usual way are solved by the adventitious aid of brazen meridians, hour indexes, &c.

My father thought that this method might be useful, to ground more deeply in the young pupil's mind, those principles which the globes are intended to explain; and by giving him a different view of the subject, improve and strengthen his mind; he therefore inserted on his globes some lines, for the purpose of solving a few problems in Lord Castlemaine's manner.

It appears to me, from a copy of Moxon's publication, which is in my possession, that the Earle of Castlemaine projected a new edition of his works, as the copy contains a great number of corrections, many alterations, and some additions. It is not very improbable, that at some future day I may re-publish this curious work, and adapt a small globe for the solution of the problems.

The meridians on our new terrestrial globes being secondaries to the equator, are also hour circles, and are marked as such with Roman figures, under the equator, and at the polar circles. But there is a difference in the figures placed to the same hour circle; if it cuts the IIId hour upon the polar circles, it will cut the IX hour upon the equator, which is six hours later, and so of all the rest.

Through the great Pacific sea, and the intersection of Libra, is drawn a broad meridian from pole to pole; it passes through the XIIth hour upon the equator, and the VIth hour upon each of the polar circles; this hour circle is graduated into degrees and parts, and numbered from the equator towards either pole.

There is another broad meridian passing through the Pacific sea, at the IXth hour upon the equator, and the IIId hour upon each polar circle; this contains only one quadrant, or 90 degrees; the numbers annexed to it begin at the northern polar circle, and end at the tropic of Capricorn.

Here we must likewise observe, there are 23 concentric circles drawn upon the terrestrial globe within the northern and southern polar circles, which for the future we shall call polar parallels; they are placed at the distance of one degree from each other, and represent the parallels of the sun's declination, but in a different manner from the 47 parallels between the tropics.

The following problems require the globe to be placed upon a plane that is level, or truly horizontal, which is easily attained, if the floor, pavement, gravel-walk in the garden, &c. should not happen to be horizontal.

A flat seasoned board, or any box which is about two feet broad, or two feet square, if the top be perfectly flat, will answer the purpose; the upper surface of either may be set truly horizontal, by the help of a pocket spirit level, or plumb rule, if you raise or depress this or that side by a wedge or two, as the spirit level shall direct; if you have a meridian line drawn on the place over which you substitute this horizontal plane, it may be readily transferred from thence to the surface just levelled; this being done, we are prepared for the solution of the following problems.

It will be necessary to define a term we are obliged to make use of in the solution of these problems, namely, the *shade of extuberancy:* by this is meant that shade which is caused by the sphericity of the globe, and answers to what we have heretofore named the terminator, defining the boundaries of the illuminated and obscure parts of the globe; this circle was, in the solution of some of the foregoing problems, represented by the broad paper circle, but is here realized by the rays of the sun.

PROBLEM XXXIV.

To observe the sun's altitude (by the terrestrial globe) when he shines bright, or when he can but just be discerned through a cloud.

Elevate the north pole of the globe to $66\frac{1}{2}$ degrees; bring that meridian, or hour circle, which passes through the IXth hour upon the equator, under the graduated side of the strong brass meridian; the globe being now set upon the horizontal plane, turn it about thereon, frame and all, that the shadow of the strong brass meridian may fall directly under itself; or in other words, that the shade of it's graduated face may fall exactly upon the aforesaid hour circle; at that instant the shade of extuberancy will touch the true degree of the sun's altitude upon that meridian, which passes through the IXth hour upon the equator, reckoned from the polar circle, the most elevated part of which will then be in the zenith of the place where this operation is performed, and is the same whether it should happen to be either in north or south latitude.

Thus we may, in an easy and natural manner, obtain the altitude of the sun, at any time of the day, by the terrestrial globe; for it is very plain, when the sun rises, he brushes the zenith and nadir of the globe by his rays; and as he always illuminates half of it, (or a few minutes more, as his globe is considerably larger than that of the earth) therefore when the sun is risen a degree higher, he must necessarily illuminate a degree beyond the zenith, and so on proportionably from time to time.

But as the illuminated part is somewhat more than half, deduct 13 minutes from the shade of extuberancy, and you have the sun's altitude with tolerable exactness.

If you have any doubt how far the shade of extuberancy reaches, hold a pin, or your finger, on the globe, between the sun and point in dispute, and where the shade of either is lost, will be the point sought.

When the sun does not shine bright enough to cast a shadow.

Turn the meridian of the globe towards the sun, as before, or direct it so that it may lie in the same plane with it, which may be done if you have but the least glimpse of the sun through a cloud; hold a string in both hands, it having first been put between the strong brass meridian and the globe; stretch it at right angles to the meridian, and apply your face near to the globe, moving your eye lower and lower, till you can but just see the sun; then bring the string held as before to this point upon the globe, that it may just obscure the sun from your sight, and the degree on the aforesaid hour circle, which the string then lies upon, will be the sun's altitude required, for his rays would shew the same point if he shone out bright.

Note. The moon's altitude may be observed by either of these methods, and the altitude of any star by the last of them.

PROBLEM XXXV.

To place the terrestrial globe in the sun's rays, that it may represent the natural position of the earth, either by a meridian line, or without it.

If you have a meridian line, set the north and south points of the broad paper circle directly over it, the north pole of the globe being elevated to the latitude of the place, and standing upon a level plane, bring the place you are in under the graduated side of the strong brass meridian, then the poles and parallel circles upon the globe will, without sensible error, correspond with those in the heavens, and each point, kingdom, and state, will be turned towards the real one which it represents.

If you have no meridian line, then the day of the month being known, find the sun's declination as before instructed, which will direct you to the parallel of the day, amongst the polar parallels, reckoned from either pole towards the polar circle; which you are to remember.

Set the globe upon your horizontal plane in the sun-shine, and put it nearly north and south by the mariner's compass, it being first elevated to the latitude of the place, and the place itself brought under the graduated side of the strong brass meridian; then move the frame and globe together, till the shade of extuberancy, or term of illumination, just touches the polar parallel for the day, and the globe will be settled as before; and if accurately performed, the variation of the magnetic needle will be shewn by the degree to which it points in the compass box.

And here observe, if the parallel for the day should not happen to fall on any one of those drawn upon the globe, you are to estimate a proportionable part between them, and reckon that the parallel of the day. If we had drawn more, the globe would have been confused.

The reason of this operation is, that as the

sun illuminates half the globe, the shade of extuberancy will constantly be 90 degrees from the point wherein the sun is vertical.

If the sun be in the equator, the shade and illumination must terminate in the poles of the world; and when he is in any other diurnal parallel, the terms of illumination must fall short of, or go beyond either pole, as many degrees as the parallel which the sun describes that day is distant from the equator; therefore, when the shade of extuberancy touches the polar parallel for the day, the artificial globe will be in the same position, with respect to the sun, as the earth really is, and will be illuminated in the same manner.

PROBLEM XXXVI.

To find naturally the sun's declination, diurnal parallel, and his place thereon.

The globe being set upon an horizontal plane, and adjusted by a meridian line or otherwise, observe upon which, or between which polar parallel the term of illumination falls; it's distance from the pole is the degree of the sun's declination; reckon this distance from the equator among the larger parallels, and you have the parallel which the sun describes that day; upon which if you move a card, cut in the form of a double square, until it's shadow falls under itself, you will obtain the very place upon that parallel over which the sun is vertical at any hour of that day, if you set the place you are in under the graduated side of the strong brass meridian.

Note, The moon's declination, diurnal parallel, and place, may be found in the same manner. Likewise, when the sun does not shine bright, his declination, &c. may be found by an application in the manner of problem xxxiv.

PROBLEM XXXVII.

To find the sun's azimuth naturally.

If a great circle, at right angles to the horizon, passes through the zenith and nadir, and also through the sun's center, it's distance from the meridian in the morning or evening of any day, reckoned upon the degrees on the inner edge of the broad paper circle, will give the azimuth required.

Method 1.

Elevate either pole to the position of a parallel sphere, by bringing the north pole in north latitude, and the south pole in south latitude, into the zenith of the broad paper circle, having first placed the globe upon your meridian line, or by the other method before prescribed; hold up a plumb line, so that it may pass freely near the outward edge of the broad paper circle, and move it so that the shadow of the string may fall upon the elevated pole; then cast your eye immediately to it's shadow on the broad paper circle, and the degree it there falls upon is the sun's azimuth at that time, which may be reckoned from either the south or north points of the horizon.

Method II.

If you have only a glimpse, or faint sight of the sun, the globe being adjusted as before, stand on the shady side, and hold the plumb line on that side also, and move it till it cuts the sun's center, and the elevated pole at the same time; then cast your eye towards the broad paper circle, and the degree it there cuts is the sun's azimuth, which must be reckoned from the opposite cardinal point.

PROBLEM XXXVIII.

To shew that in some places of the earth's surface, the sun will be twice in the same azimuth in the morning, twice in the same azimuth in the afternoon: or in other words,

When the declination of the sun exceeds the latitude of any place, on either side of the

OF THE GLOBES. 107

equator, the sun will be on the same azimuth twice in the morning, and twice in the afternoon.

Thus, suppose the globe rectified to the latitude of Antigua, which is about 17 deg. of north latitude, and the sun to be in the beginning of Cancer, or to have the greatest north declination; set the quadrant of altitude to the 21st degree north of the east in the horizon, and turn the globe upon it's axis, the sun's center will be on that azimuth at 6 h. 30 min. and also at 10 h. 30 min. in the morning. At 8 h. 30 min. the sun will be as it were stationary, with respect to it's azimuth, for some time; as it will appear by placing the quadrant of altitude to the 17th degree north of the east in the horizon. If the quadrant be set to the same degrees north of the west, the sun's center will cross it twice as it approaches the horizon in the afternoon.

This appearance will happen more or less to all places situated in the torrid zone, whenever the sun's declination exceeds their latitude; and from hence we may infer, that the shadow of a dial, whose gnomon is erected perpendicular to an horizontal plane, must necessarily go back several degrees on the same day.

But as this can only happen within the torrid zone, and as Jerusalem lies about 8 degrees to the north of the tropic of Cancer, the retrocession of the shadow on the dial of Ahaz, at Jerusalem, was, in the strictest signification of the word, miraculous.

PROBLEM XXXIX.

To observe the hour of the day in the most natural manner, when the terrestrial globe is properly placed in the sun-shine.

There are many ways to perform this operation with respect to the hour, three of which are here inserted, being general to all the inhabitants of the earth; a fourth is added, peculiar to those of London, which will answer, without sensible error, at any place not exceeding the distance of 60 miles from this capital.

1st, By a natural style.

Having rectified the globe as before directed, and placed it upon an horizontal plane over your meridian line, or by the other method, hold a long pin upon the illuminated pole, in the direction of the polar axis, and it's shadow will shew the hour of the day amongst the polar parallels.

The axis of the globe being the common section of the hour circles, is in the plane of each; and as we suppose the globe to be properly adjusted, they will correspond with those in the heavens; therefore the shade of a pin, which is the axis continued, must fall upon the true hour circle.

2dly, *By an artificial stile.*

Tie a small string, with a noose, round the elevated pole, stretch it's other end beyond the globe, and move it so that the shadow of the string may fall upon the depressed axis; at that instant it's shadow upon the equator will give the solar hour to a minute.

But remember, that either the autumnal or vernal equinoctial colure must first be placed under the graduated side of the strong brass meridian, before you observe the hour, each of these being marked upon the equator with the hour XII.

The string in this last case being moved into the plane of the sun, corresponds with the true hour circle, and consequently gives the true hour.

3dly, *Without any stile at all.*

Every thing being rectified as before, look where the shade of extuberancy cuts the equator, the colure being under the graduated side of the strong brass meridian, and you obtain the hour in two places upon the equator, one of them going before, and the other following the sun.

Note, If this shade be dubious, apply a pin, or your finger, as before directed.

The reason is, that the shade of extuberancy being a great circle, cuts the equator in half, and the sun, in whatsoever parallel of declination he may happen to be, is always in the pole of the shade; consequently the confines of light and shade will shew the true hour of the day.

4*thly, Peculiar to the inhabitants of London, and any place within the distance of sixty miles from it.*

The globe being every way adjusted as before, and London brought under the graduated side of the strong brass meridian, hold up a plumb-line, so that it's shadow may fall upon the zenith point, (which in this case is London itself) and the shadow of the string will cut the parallel of the day upon that point to which the sun is then vertical, and that hour circle upon which this intersection falls, is the hour of the day; and as the meridians are drawn within the tropics, at twenty minutes distance from each other, the point cut by the intersection of the string upon the parallel of the day, being so near the equator, may, by a glance of the observer's eye, be referred thereto, and the true time obtained to a minute.

The plumb-line thus moved is the azimuth; which, by cutting the parallel of the day, gives the sun's place, and consequently the hour circle which intersects it.

From this last operation results a corollary, that gives a second way of rectifying the globe to the sun's rays.

If the azimuth and shade of the illuminated axis agree in the hour when the globe is rectified, then making them thus to agree, must rectify the globe.

COROLLARY.

Another method to rectify the globe to the sun's rays.

Move the globe, till the shadow of the plumb-line, which passes through the zenith cuts the same hour on the parallel of the day, that the shade of the pin, held in the direction of the axis, falls upon, amongst the polar parallels, and the globe is rectified.

The reason is, that the shadow of the axis represents an hour circle; and by it's agreement in the same hour, which the shadow of the azimuth string points out, by it's intersection on the parallel of the day, it shews the sun to be in the plane of the said parallel; which can never happen in the morning on the eastern side of the globe, nor in the evening on the western side of it, but when the globe is rectified.

This rectification of the globe is only placing it in such a manner, that the principal great circles and points may concur and fall in with those of the heavens.

The many advantages arising from these problems, relating to the placing of the globe in the sun's rays, the tutor will easily discern, and readily extend to his own, as well as to the benefit of his pupil.

THE

GENERAL PRINCIPLES

OF

DIALLING

ILLUSTRATED

BY THE TERRESTRIAL GLOBE.

THE art of dialling is of very ancient origin, and was in former times cultivated by all who had any pretensions to science; and before the invention of clocks and watches it was of the highest importance, and is even now used to correct and regulate them.

It teaches us, by means of the sun's rays, to divide time into equal parts, and to repre-

OF THE GLOBES. 113

sent on any given surface the different circles into which, for convenience, we suppose the heavens to be divided, but principally the hour circles.

The hours are marked upon a plane, and pointed out by the interposition of a body which receiving the light of the sun, casts a shadow upon the plane. This body is called the axis, when it is parallel to the axis of the world. It is called the stile, when it is so placed that only the end of it coincides with the axis of the earth; in this case, it is only this point which marks the hours.

Among the various pleasing and profitable amusements which arise from the use of globes, that of dialling is not the least. By it the pupil will gain satisfactory ideas of the principles on which this branch of science is founded; and it will reward, with abundance of pleasure, those that chuse to exercise themselves in the practice of it.

If we imagine the hour circles of any place, as London, to be drawn upon the globe of the earth, and suppose this globe to be transparent, and to revolve round a real axis, which is opake, and casts a shadow; it is evident, that whenever the plane of any hour semicircle points at the sun, the shadow of the axis will fall upon the opposite semicircle.*

* Long's Astronomy, vol i, page 82.

Let a P C p, fig. 1, plate XIII, represent a transparent globe; a b c d e f g the hour semicircles; it is clear, that if the semicircle P a p points at the sun, the shadow of the axis will fall upon the opposite semicircle.

If we imagine any plane to pass through the center of this transparent globe, the shadow of half the axis will always fall upon one side or the other of this intersecting plane.

Thus let A B C D be the plane of the horizon of London; so long as the sun is above the horizon, the shadow of the upper half of the axis will fall somewhere upon the upper side of the plane A B C D; when the sun is below the horizon of London, then the shadow of the lower half of the axis E falls upon the lower side of the plane.

When the plane of any hour semicircle points at the sun, the shadow of the axis marks the respective hour-line upon the intersecting plane. The hour-line is therefore a line drawn from the center of the intersecting plane, to that point where this plane is cut by the semicircle opposite to the hour semicircle.

Thus let A B C D, fig. 1, plate XIII, the horizon of London, be the intersecting plane; when the meridian of London points at the sun, as in the present figure, the shadow of the half axis P E falls upon the line E B, which is drawn from E, the center of the horizon, to the point where the horizon is cut by the opposite semicircle; therefore, E B is the line for the hour of twelve at noon.

By the same method the rest of the hour-lines are found, by drawing for every hour a line, from the center of the intersecting plane, to that semicircle which is opposite to the hour semicircle.

Thus fig. 2, plate XIII, shews the hour-lines drawn upon the plane of the horizon of London, with only so many hours as are necessary; that is, those hours, during which the sun is above the horizon of London, on the longest day in summer.

If, when the hour-lines are thus found, the semicircles be taken away, as the scaffolding is when the house is built, what remains, as in fig. 2, will be an *horizontal dial* for London.

If, instead of twelve hour circles, as above described, we take twice that number, we may by the points, where the intersecting plane is cut by them, find the lines for every half hour; if we take four times the number of hour circles, we may find the lines for every quarter of an hour, and so on progressively.

We have hitherto considered the horizon of London as the intersecting plane, by which is seen the method of making an horizontal dial. If we take any other plane for the intersecting plane, and find the points where the hour semicircles pass through it, and draw the lines from the center of the plane to those points, we shall have the hour-lines for that plane.

Fig. 3, plate XIII, shews how the hour-lines are found upon a south plane, perpendicular to the horizon. Fig. 4, shews a south dial, with it's hour-lines, without the semicircle, by means whereof they are found.

The *gnomon* of every sun-dial represents the axis of the earth, and is therefore always placed parallel to it; whether it be a wire, as in the figure before us, or the edge of a brass plate, as in a common horizontal dial.

The whole earth, as to it's bulk, is but a *point*, if compared to it's distance from the sun; therefore, if a small sphere of glass be placed on any part of the earth's surface, so that it's axis be parallel to the axis of the earth, and the sphere have such lines upon it, and such planes within it, as above described, it will shew the hour of the day as truly as if it were placed at the center of the earth, and the shell of the earth were as transparent as glass.

A wire sphere, with a thin flat plate of brass within it, is often made use of to explain the principles of dialling.

From what has been said, it is clear that dialling depends on finding where the shadow of a strait wire, parallel to the axis of the earth, will fall upon a given plane, every hour, half hour, &c. the hour-lines being found as above described, which we shall proceed to exemplify by the globe.

Every dial-plane (that is, the plane surface on which a dial is drawn) represents the plane of a great circle, which circle is an *horizon* to some country or other.

The center of the dial represents the center of the earth; and the gnomon which casts the shade represents the axis, and ought to point directly to the poles of the equator.

The plane upon which dials are delineated may be either, 1. parallel to the horizon; 2. perpendicular to the horizon; or, 3. cutting it at oblique angles.

PROBLEM. XL.

To construct an horizontal dial for any given latitude, by means of the terrestrial globe.

Elevate the globe to the latitude of the place, then bring the first meridian under the graduated edge of the strong brazen one, which will then be over the hour XII, or the equator. As our globes have meridians drawn through every fifteen degrees of the equator, these meridians will represent the true circles of the sphere, and will intersect the horizon of the globe, in certain points on each side of the meridian. The distance of these points from the meridian must be carefully noted down upon a piece of paper, as will be seen in the example. The pupil need not, however, take out into his table the distances further than from XII to VI, which is just 90 degrees; for the distances of XI, X, IX, VIII, VII, VI, in the forenoon, are the same from XII as the distances of I, II, III, IV, V, VI, in the afternoon; and these hour-lines continued through the center will give the opposite hour-lines on the other half of the dial.

No more hour-lines need be drawn than what answer to the sun's continuance above the horizon, on the longest day of the year, in the given latitude.

Example. Suppose the given place to be London, whose latitude is 51 deg. 30 min. north.

Elevate the north pole of the globe to $51\frac{1}{2}$ degrees above the horizon; then will the axis of the globe have the same elevation above the broad paper circle, as the gnomon of the dial is to have above the plane thereof.

Turn the globe, till the first meridian (which on English globes passes through London) is under the graduated side of the strong brazen meridian; then observe and note the points where the hour-circles intersect the horizon; and as on our globes the inner graduated circle, on the broad paper circle, begins from the two sixes, or east and west, we shall begin from thence,

OF THE GLOBES. 119

calling the hour - - - VI 0° 0
we shall find the other hours intersecting the
horizon at the following degrees: V 18° 54
IV 36 24
III 51 57
II 65 41
I 78 9
which are the respective distances of the above
hours from VI upon the plane of the horizon.

To transfer these, and the rest of the hours, upon an horizontal plane, draw the parallel right lines a c and b d, fig. 5, plate XIII, upon that plane, as far from each other as is equal to the intended thickness of the gnomon of the dial, and the space included between them will be the meridian, or twelve o'clock line upon the dial; cross this meridian at right angles by the line g h, which will be the six o'clock line; then setting one foot of your compasses in the intersection a, describe the quadrant g e with any convenient radius, or opening of the compasses; after this, set one foot of the compasses in the intersection b, as a center, and with the same radius describe the quadrant f h; then divide each quadrant into 90 equal parts, or degrees, as in the figure.

Because the hour-lines are less distant from each other about noon, than in any other part of the dial, it is best to have the centers of the quadrants at some distance from the center of the dial-plane, in order to enlarge the hour-distances near XII; thus the center of the plane is at A, but the center of the quadrants is at a and b.

Lay a rule over 78° 9', and the center b, and draw there the hour-line of I. Through b, and 65 41, gives the hour-line of II. Through b, and 51 57, that of III. Through the same center, and 36 24, we obtain the hour-line of IV. And through it, and 18 54, that of V. And because the sun rises about four in the morning, continue the hour-lines of IV and V in the afternoon, through the center b to the opposite side of the dial.

Now lay a rule successively to the center a of the quadrant e g, and the like elevations or degrees of that quadrant, 78 9, 65 41, 51 57, 36 24, 18 54, which will give the forenoon hours of XI, X, IX, VIII, and VII; and because the sun does not set before VIII in the evening on the longest days, continue the hour-lines of VII and VIII in the afternoon, and all the hour lines will be finished on this dial.

Lastly, through $51\frac{1}{2}$ degrees on either quadrant, and from it's center, draw the right line a g for the axis of the gnomon a g i, and from g let fall the perpendicular g i upon the meridian line a i, and there will be a triangle made, whose sides are a g, g i, and i a; if a plate similar to this triangle be made as thick as the distance between the lines a c and b d, and be set upright between them, touching at a and b, the line a g will, when it is truly set, be parallel to the axis of the world, and will cast a shadow on the hour of the day.

The trouble of dividing the two quadrants may be saved, by using a line of chords, which is always placed upon every scale belonging to a case of instruments.

PROBLEM XLI.

To delineate a direct south dial for any given latitude, by the globe.

Let us suppose a south dial for the latitude of London.

Elevate the pole to the co-latitude of your place, and proceed in all respects as above taught for the horizontal dial, from VI in the morning to VI in the afternoon, only the hours must be reversed, as in fig. 3, plate XIII; and the hypothenuse a g of the gnomon a g f, must make an angle with the dial plane to the co-latitude of the place.

As the sun can shine no longer than from VI in the morning to VI in the evening, there is no occasion for having more than twelve hours upon this dial.

In solving this problem, we have considered our vertical south dial for the latitude of London, as an horizontal one for the complement of that latitude, or 38 deg. 30 min.; all direct vertical dials may be thus reduced to horizontal ones, in the same manner. The reason of this will be evident, if the globe be elevated to the latitude of London; for by fixing the quadrant of altitude to the zenith, and bringing it to intersect the horizon in the east point, it will point out the plane of the proposed dial.

This plane is at right angles to the meridian, and perpendicular to the horizon; and it is clear, from the bare inspection of the globe thus elevated, that it's axis forms an angle with this plane, which is just the complement of that which it forms with the horizon, and is therefore just equal to the co-latitude of the place; and that therefore it is most simple to rectify the globe to that co-latitude.

The north vertical dial is the same with the south, only the stile must point upwards, and that many of the hours from it's direction can be of no use.

PROBLEM XLII.

To make an erect dial, declining from the south towards the east or west.

Elevate the pole to the latitude of the place, and screw the quadrant of altitude to the zenith.

OF THE GLOBES. 123

Then if your dial declines towards the east, (which we shall suppose in the present instance) count in the horizon the degrees of declination from the east point towards the north, and bring the lower end of the quadrant to coincide with that degree of declination at which the reckoning ends.

Then bring the first meridian under the graduated edge of the strong brass meridian, which strong meridian will be the horary index.

Now turn the globe westward, and observe the degrees cut in the quadrant of altitude by the first meridian, while the hours XI, X, IX, &c. in the forenoon, pass successively under the brazen one; and the degrees thus cut on the quadrant by the first meridian, are the respective distances of the forenoon hours, from XII, on the plane of the quadrant.

For the afternoon hours, turn the quadrant of altitude round the zenith, until it comes to the degree in the horizon, opposite to that where it was placed before, namely, as far from the west towards the south, and turn the globe eastward; and as the hours I, II, III, &c. pass under the strong brazen meridian, the first meridian will cut on the quadrant of altitude the number of degrees from the zenith, that each of the hours is from XII on the dial.

When the first meridian goes off the quadrant at the horizon, in the forenoon, the hour index will shew the time when the sun comes upon this dial; and when it goes off the quadrant in the afternoon, it points to the time when the sun leaves the dial.

Having thus found all the hour distances from XII, lay them down upon your dial plane, either by dividing a semicircle into two quadrants, or by the line of chords.

In all declining dials, the line on which the gnomon stands makes an angle with the twelve o'clock line and falls among the forenoon hour lines, if the dial declines towards the east; and among the afternoon hour lines, when the dial declines towards the west; that is, to the left hand from the twelve o'clock line in the former case, and to the right hand from it in the latter.

To find the distance of this line from that of twelve.

This may be considered, 1. If the dial declines from the south towards the east, then count the degrees of that declination in the horizon, from the east point towards the north, and bring the lower end of the quadrant to that degree of declination where the reckoning ends; then turn the globe, until the first meridian cuts the horizon in the like number of degrees, counted from the south point towards the east, and the quadrant and first meridian will cross one another at right angles, and the number of degrees of the quadrant, which are intercepted between the first meridian and the zenith, is equal to the distance of this line from the twelve o'clock line.

The numbers of the first meridian, which are intercepted between the quadrant and the north pole, is equal to the elevation of the stile above the plane of the dial.

The second case is, when the dial declines westward from the south.

Count the declination from east point of the horizon, towards the south, and bring the quadrant of altitude to the degree in the horizon, at which the reckoning ends, both for finding the forenoon hours, and the distance of the substile, or gnomon line, from the meridian; and for the afternoon hours, bring the quadrant to the opposite degrees in the horizon, namely, as far from the west towards the north, and then proceed in all respects as before.

It is presumed, that the foregoing instances will be sufficient to illustrate the general principles of dialling, and to give the pupil a general idea of that pleasing science; for accurate and expeditious methods of constructing dials, we must refer him to treatises written expressly on that subject.

NAVIGATION

EXPLAINED BY THE GLOBE.

NAVIGATION is the art of guiding a ship at sea, from one place to another, in the safest and most convenient manner. In order to attain this, four things are particularly necessary:

1. To know the situation and distance of places.

2. To know at all times the points of the compass.

3. To know the line which the ship is to be directed from one place to the other.

4. To know, in any part of the voyage, what point of the globe the ship is upon.

The knowledge of the distance and situation of places, between which a voyage is to be made, implies not only a general knowledge of geography, but of several other particulars, as the rocks, sands, streights, rivers, &c. near which we are to sail; the bending out, or running in of the shores, the knowledge of the times that particular winds sets in, the seasons when storms and hurricanes are to be expected, but especially the tides; these and many other similar circumstances are to be learned from sea charts, journals, &c. but chiefly by observation and experience.

The second particular to be attained, is the knowledge at all times of the points of the compass, where the ship is. The ancients, to whom the polarity of the loadstone was unknown, found in the day-time the east or west, by the rising or setting of the sun; and at night, the north by the polar star. We have the advantage of the mariner's compass, by which, at any time in the wide ocean, and the darkest night, we know where the north is, and consequently the rest of the points of the compass.

Indeed, before the invention of the mariner's compass, the voyages of the Europeans were principally confined to coasting; but this fortunate discovery has enabled the mariner to explore new seas, and discover new countries, which, without this valuable acquisition, would probably have remained for ever unknown.

The third thing required to be known, is the line which a ship describes upon the globe of the earth, in going from one place to another.

The shortest way from one place to another, is an arc of a great circle, drawn through the two places.

The most convenient way for a ship, is that by which we may sail from one place to another, directing the ship all the while towards the same point of the compass.

A ship is guided by steering or directing her towards some points of the compass; the line wherein a ship is directed, is called the ship's course, which is named from the point towards which she sails.

Thus if a ship sails towards the north-east point, her course is said to be N. E.

In long voyages, a ship's way may consist of a great number of different courses, as from A to B, from B to C, and from C to D, fig 9, plate XIII; when we speak of a ship's course, we consider one of these at a time; the seldomer the course is changed, the more easily the ship is directed.

If two places, A *and* Z, *fig.* 7, *plate* XIII. *lie under the same meridian,* the course from the one side to the other is due north or south. Thus let A Z be part of a meridian; if A be south of Z, the course from A to Z must be north, and the course from Z to A south. This is evident from the nature of a meridian, that it marks upon the horizon the north and south points, and that every point of any meridian is north or south from every other point of it. From hence we may deduce the following corollary; that if a ship sails due north or south, she will continue on the same meridian.

If two places lie under the equator, the course from one to the other is an arc of the equator, and is due east or west. Thus let a z, fig. 7, be a part of the equator; if a be west from z, the course from a to z is east, and the course from z to a is west: for since the equator marks the east and west points upon the horizon, every point of the equator lies east or west of every other point of it, as may be seen upon the globe, by placing it as for a right sphere, and bringing a or z, or any of the intermediate points, to the zenith; when it will be evident, that if we are to go from one of these points a, to the other z, or to any point on the equator, we must continue our course due east to arrive at a, or vice versa. From hence we may deduce this consequence, that if a ship under the equator sails due east or west, she will continue under the equator.

In the two foregoing cases, the course being an arc of a great circle, (the meridian or equator) is the shortest and the most convenient way it can sail.

If two places lie under the same parallel, the course from one to the other is due east or west; this may be seen upon the globe, by the following method: bring any point of a parallel to the zenith, and stretch a thread over it, perpendicular to the meridian; the thread will then be a tangent to the parallel, and stand east and west from the point of contact. Hence, If a ship sails in any parallel, due east or west, she will continue in the same parallel. In this case, the most convenient course, though not the shortest, from one to the other, is to sail due east or west.

If two places lie neither under the equator, nor on the same meridian, nor in the same parallel, the most convenient, though not the shortest, course from one to the other, is in a rhumb.

For if we should in this case attempt to go the shortest way, in a great circle drawn through the two places, we must be perpetually changing our course. Thus fig. 8, whatever is the bearing of Z from A, the bearings of all the intermediate points, as B, C, D, E, &c. will be different from it, as well as different from each other, as may be easily seen upon the globe, by bringing the first point A to the zenith, and observing the bearing of Z from each of them. Thus suppose, when the globe is rectified to the horizon of A, the bearing of Z from A be northeast, and the angle of position of Z, with regard to A, be 45 degrees; if we bring B to the zenith, we shall have a different horizon, and the bearing and angle of position from Z to B will be different from the former; and so on of the other points C, D, E, they will each of them have a different horizon, and Z will have a different bearing and angle of position.

From hence we may draw this corollary, that when two places lie one from the other, towards a point not cardinal, if we sail from one place towards the point of the other's bearing, we shall never arrive at the other place. Thus if Z lies north-east from A, if we sail from A towards the north-east, we shall never arrive at Z.

A *rhumb* upon the globe is a line drawn from a given place A, so as to cut all the meridians it passes through at equal angles; the rhumbs are denominated from the points of the compass, in a different manner from the winds. Thus, at sea, the north-east wind is that which blows from the north-east point of the horizon, towards the ship in which we are; but we are said to sail upon the N. E. rhumb, when we go towards the north-east.

The rhumb A B C D Z, fig. 8, plate XIII. passing through the meridians L M, N O, P Q, makes the angles L A B, N B C, P C D, equal; from whence it follows, that the direction of a rhumb is in every part of it towards the same point of the compass; thus from every point of a north-east rhumb upon the globe, the direction is towards the north-east, and that rhumb makes an angle of 45 deg. with every meridian it is drawn through.

Another property of the rhumbs is, that equal parts of the same rhumb are contained between parallels of equal distance of latitude; so that a ship continuing in the same rhumb, will run the same number of miles in sailing from the parallel of 10 to the parallel of 30, as she does in sailing from the parallel of 30 to that of 50.

The fourth thing mentioned to be required in navigation, was, to know at any time what point of the globe a ship is upon. This depends upon four things: 1. the longitude; 2. the latitude; 3. the course the ship has run; 4. the distance, that is, the way she has made, or the number of leagues or miles she has run in that course, from the place of the last observation. Now any two of these being known, the rest may be easily found.

Having thus given some general idea of navigation, we now proceed to the problems by which the cases of sailing are solved on the globe.

PROBLEM XLIII.

*Given the difference of latitude, and difference of longitude, to find the course and distance sailed.**

Example. Admit a ship sails from a port

---

* See Martin on the Globes.

A, in latitude 38 deg. to another port B, in latitude 5 deg. and finds her difference of longitude 43 deg.

Let the port A be brought to the meridian, and elevate the globe to the given latitude of that port 38 deg. and fixing the quadrant of altitude precisely over it on the meridian, move the quadrant to lie over the second port B, (found by the given difference of latitude and longitude) then will it cut in the horizon 50 deg. 45 min. for the angle of the *ship's course* to be steered from the port A. Also, count the degrees in the quadrant between the two ports, which you will find 51 deg.; this number multiplied by 60, (the nautical miles in a degree) will give 3060 for the distance run.

PROBLEM XLIV.

Given the difference of latitude and course, to find the difference of longitude and distance sailed.

Example. Admit a ship sails from a port A, in 25 deg. north latitude, to another port B, in 30 deg. south latitude, upon a course of 43 deg.

Bring the port A to the meridian, and rectify the globe to the latitude thereof 25 deg. where fix the quadrant of altitude, and place it so as to make an angle with the meridian of

43 deg. in the horizon, and observe where the edge of the quadrant intersects the parallel of 30 deg. south latitude, for that is the place of the port B. Then count the number of degrees on the edge of the quadrant intersected between the two ports, and there will be found 73 deg. which, multiplied by 60, gives 4380 miles for the distance sailed. As the two ports are now known, let each be brought to the meridian, and observe the difference of longitude in the equator respectively, which will be found 50 degrees.

N. B. Had this problem been solved by *loxodromics*, or sailing on a rhumb, the difference of longitude would then have been 52 deg. 30 min. between the two ports.

PROBLEM XLV.

Given the difference of latitude and distance run, to find the difference of longitude, and angle of the course.

Example. Admit a ship sails from a port A, in latitude 50 deg. to another port B, in latitude 17 deg. 30 min. and her distance run be 2220 miles. Rectify the globe to the latitude of the place A, then the distance run, reduced to degrees, will make 37 deg. which are to be reckoned from the end of the quadrant lying over the port A, under the meridian;

then is the quadrant to be moved, till the 37 deg. coincides with the parallel of 17 deg. 30 min. north latitude; then will the angle of the course appear in the arch of the horizon, intercepted between the quadrant and the meridian, which will be 32 deg. 40 min.; and by making a mark on the globe for the port B, and bringing the same to the meridian, you will observe what number of degrees pass under the meridian, which will be 20, the difference of longitude required.

PROBLEM XLVI.

Given the difference of longitude and course, to find the difference of latitude and distance sailed.

Example. Suppose a ship sails from A, in the latitude 51 deg. on a course making an angle with the meridian of 40 deg. till the difference of longitude be found just 20 deg.; then rectifying the globe to the latitude of the port A, place the quadrant of altitude so as to make an angle of 40 deg. with the meridian; then observe at what point it intersects the meridian passing through the given longitude of the port B, and there make a mark to represent the said port; then the number of degrees intercepted between that and the port A will be 28, which will give 1680 miles for the distance run. And the said mark for the port B, being brought to the meridian, will have it's latitude there shewn to be 27 deg. 40 min.

PROBLEM XLVII.

Given the course and distance sailed, to find the difference of longitude, and difference of latitude.

Example. Suppose a ship sails 1800 miles from a port A, 51 deg. 15 min, south-west, on an angle of 45 deg. to another port B.

Having rectified the globe to the port A, fix the quadrant of altitude over it in the zenith, and place it to the south-west point in the horizon; then upon the edge of the quadrant under 30 deg. (equal to 1800 miles from the port A) is the port B; which bring to the meridian, and you will there see the latitude; and at the same time, it's longitude on the equator, in the point cut by the meridian.

In all these cases, the ship is supposed to be kept upon the *arch of a great circle*, which is not difficult to be done, very nearly, by means of the globe, by frequently observing the latitude, measuring the distance sailed, and (when you can) finding the difference of longitude; for one of these being given, the place and course of the ship is known at the same time; and therefore the preceding course may be altered, and rectified without any trouble, through the whole voyage, as often as such observations can be obtained, or it is found necessary. Now if any of these *data* are but of the quantity of four or five degrees, it will suffice for correcting the ship's course by the globe, and carrying her directly to the intended port, according to the following problem.

PROBLEM XLVIII.

To steer a ship upon the arch of a great circle by the given difference of latitude, or difference of longitude, or distance sailed in a given time.

Admit a ship sails from a port A, to a very distant port Z, whose latitude and longitude are given, as well as it's geographical bearing from A; then,

First, having rectified the globe to the port A, lay the quadrant of altitude over the port Z, and draw thereby the arch of the great circle through A and Z; this will design the intended path or tract of the ship.

Secondly, having kept the ship upon the first given course for some time, suppose by an observation you find the latitude of the *present place* of the ship, this added to, or subducted from the latitude of the port A, will give the present latitude in the meridian; to which bring the path of the ship, and the part therein, which lies under the new latitude, is the true place B of the ship in the great arch. To the latitude of B rectify the globe, and lay the quadrant over Z, and it will shew in the horizon the new course to be steered.

Thirdly, suppose the ship to be steered upon this course, till her distance run be found 300 miles, or 5 deg.; then, the globe being rectified to the place B in the zenith, laying the quadrant from thence over the great arch, make a mark at the 5th degree from B, and that will be the present place of the ship, which call C; which being brought to the meridian, it's latitude and longitude will be known. Then rectify the globe to the place C, and laying the quadrant from thence to Z, the new course to be steered will appear in the horizon.

Fourthly, having steered some time upon this new course, suppose, by some means or other, you come to know the difference of longitude of the present place of the ship, and of any of the preceding places, C, B, A; as B, for instance; then bring B to the meridian, and turn the globe about, till so many degrees of the equator pass under the meridian as are equal to the discovered distance of longitude; then the point of the great arch cut by the meridian is the present place D of the ship, to which the new course is to be found as before.

And thus, by repeating these observations at proper intervals, you will find future places, E, F, G, &c. in the great arch; and by rectifying the course at each, your ship will be conducted on the great circle, or the nearest way from the port A to Z, by the *use of the globe* only.

OF THE USE

OF THE

TERRESTRIAL GLOBE,

WHEN MOUNTED

IN THE COMMON MANNER.

ALTHOUGH I have, in the first part of this essay, laid before my readers the reasons which induce me to prefer my father's manner of mounting the globes, to the old or Ptolemaic form, yet as many may be in possession of globes mounted in the old form, and others may have been taught by those globes, I thought it would render these essays more complete, to give an account of so many of the leading problems, solved on the common globes, as would enable them to apply the remainder of those heretofore solved, to their own use. This is the more expedient, as, since the publication of my father's treatise, there have been a few attempts to do away some of the inconveniences of the ancient form, particular that of the old hour-circle, which is now generally placed under the meridian.

I cannot, however, refrain from again observing to the pupil, that the solution of the problems on the old globes depends upon appearances; that therefore, if he means to content himself with the mere mechanical solution of them, the Ptolemaic globes will answer his purpose; but if he wishes to have clear ideas of the rationale of those problems, he must use those mounted in my father's manner.

The celestial globe is used the same way in both mountings, excepting that in my father's mounting it has some additional circles; but the difference is so trifling, that it is presumed the pupil can find no difficulty in applying the directions there given to the old form.

PROBLEM I.

To find the latitude and longitude of any given place on the globe.

Bring the place to the east side of the brass meridian, then the degree marked on the meridian over it shews it's latitude, and the degree of the equator under the meridian shews it's longitude.

Hence, if the longitude and latitude of any place be given, the place is easily found, by bringing the given longitude to the meridian; for then the place will lie under the given degree of latitude upon the meridian.

PROBLEM II.

To find the difference of longitude between any two given places.

Bring each of the given places successively to the brazen meridian, and see where the meridian cuts the equator each time; the number of degrees contained between those two points, if it be less than 180 deg. otherwise the remainder to 360 deg. will be the difference of longitude required.

PROBLEM III.

To rectify the globe for the latitude, zenith, and sun's place.

Find the latitude of the place by prob. 1, and if the place be in the northern hemisphere, elevate the north pole above the horizon, according to the latitude of the place. If the place be in the southern hemisphere, elevate the south pole above the south point of the horizon, as many degrees as are equal to the latitude.

Having elevated the globe according to it's latitude, count the degrees thereof upon the meridian from the equator towards the elevated pole, and that point will be the zenith, or the vertex of the place; to this point of the meridian fasten the quadrant of altitude, so that the graduated edge thereof may be joined to the said point.

Having brought the sun's place in the ecliptic to the meridian, set the hour index to twelve at noon, and the globe will be rectified to the sun's place.

PROBLEM IV.

The hour of the day at any place being given, to find all those on the globe, where it is noon, midnight, or any given hour at that time.

On the globes when mounted in the common manner, it is now customary to place the hour-circle under the north pole; it is divided into twice twelve hours, and has two rows of figures, one running from east to west, the other from west to east; this circle is moveable, and the meridian answers the purpose of an index.

Bring the given place to the brazen meridian, and the given hour of the day on the hour-circle, this done, turn the globe about, till the meridian points at the hour desired; then, with all those under the meridian, it is noon, midnight, or any given hour at that time.

PROBLEM V.

The hour of the day at any place being given, to find the corresponding hour (or what o'clock it is at that time) in any other place.

Bring the given place to the brazen meridian, and set the hour-circle to the given time; then turn the globe about, until the place where the hour is required comes to the meridian, and the meridian will point out the hour of the day at that place.

Thus, when is is noon at London, it is

		H. M.	
At {	Rome - - -	0 52	P. M.
	Constantinople - -	2 7	P. M.
	Vera Cruz - - -	5 30	A. M.
	Pekin in China - -	7 50	P. M.

PROBLEM VI.

The day of the month being given, to find all those places on the globe where the sun will be vertical, or in the zenith, that day.

Having found the sun's place in the ecliptic for the given day, bring the same to the brazen meridian, observe what degree of the meridian is over it, then turn the globe round it's axis, and all places that pass under that degree of the meridian, will have the sun vertical, or in the zenith, that day; *i. e.* directly over the head of each place at it's respective noon.

PROBLEM VII.

A place being given in the torrid zone, to find those two days in the year on which the sun will be vertical to that place.

Bring the given place to the brazen meridian, and mark the degree of latitude that is exactly over it on the meridian; then turn the globe about it's axis, and observe the two points of the ecliptic which pass exactly under that degree of latitude, and look on the horizon for the two days of the year in which the sun is in those points or degrees of the ecliptic, and they are the days required; for on them, and none else, the sun's declination is equal to the latitude of the given place.

PROBLEM. VIII.

To find the antœci, periœci, and antipodes of any given place.

Bring the given place to the brazen meridian, and having found it's latitude, keep the globe in that position, and count the same number of degrees of latitude on the meridian, from the equator towards the contrary pole, and where the reckoning ends, that will give the place of the antœci upon the globe. Those who live at the equator have no antœci.

The globe remaining in the same position, bring the upper XII on the horary circle to the meridian, then turn the globe about till the meridian points to the lower XII; the place which then lies under the meridian, having the same latitude with the given place, is the periœci required. Those who live at the poles, if any, have no periœci.

As the globe now stands (with the index at the lower XII), the antipodes of the given place are under the same point of the brazen meridian where it's antœci stood before.

PROBLEM. IX.

To find at what hour the sun rises and sets any day in the year, and also upon what point of the compass.

Rectify the globe for the latitude of the place you are in; bring the sun's place to the meridian, and bring the XII to the meridian; then turn the sun's place to the eastern edge of the horizon, and the meridian will point out the hour of rising; if you bring it to the western edge of the horizon, it will shew the setting.

Thus on the 16th day of March, the sun rose a little past six, and set a little before six.

Note. In the summer the sun rises and sets a little to the northward of the east and west points, but in winter, a little to the southward of them. If, therefore, when the sun's place is brought to the eastern and western edges of the horizon, you look on the inner circle, right against the sun's place, you will see the point of the compass upon which the sun rises and sets that day.

PROBLEM. X.

To find the length of the day and night at any time of the year.

Only double the time of the sun's rising that day, and it gives the length of the night; double the time of his setting, and it gives the length of the day.

This problem shews how long the sun stays with us any day, and how long he is absent from us any night.

Thus on the 26th of May the sun rises about four, and sets about eight; therefore the day is sixteen hours long, and the night eight.

PROBLEM XI.

To find the length of the longest or shortest day, at any place upon the earth.

Rectify the globe for that place, bring the beginning of Cancer to the meridian, bring XII to the meridian, then bring the same degree of Cancer to the east part of the horizon, and the meridian will shew the time of the sun's rising.

If the same degree be brought to the western side, the meridian will shew the setting, which doubled, (as in the last problem) will give the length of the longest day and shortest night.

If we bring the beginning of Capricorn to the meridian, and proceed in all respects as before, we shall have the length of the longest night and shortest day.

Thus in the Great Mogul's dominions, the longest day is fourteen hours, and the shortest night ten hours. The shortest day is ten hours, and the longest night fourteen hours.

At Petersburgh, the seat of the Empress of Russia, the longest day is about $19\frac{1}{2}$ hours, and the shortest night $4\frac{1}{2}$ hours; the shortest day $4\frac{1}{2}$ hours, and longest night $19\frac{1}{2}$ hours.

Note. In all places near the equator, the sun rises and sets at six the year round. From thence to the polar circles, the days increase as the latitude increases; so that at those circles themselves, the longest day is 24 hours, and the longest night just the same. From the polar circles to the poles, the days continue to lengthen into weeks and months; so that at the very pole, the sun shines for six months together in summer, and is absent from it six months in winter.

Note. That when it is summer with the northern inhabitants, it is winter with the southern, and the contrary; and every part of the world partakes of an equal share of light and darkness.

OF THE GLOBES. 149

PROBLEM XII.

To find all those inhabitants to whom the sun is this moment rising or setting, in their meridian or midnight.

Find the sun's place in the ecliptic, and raise the pole as much above the horizon as the sun (that day) declines from the equator; then bring the place where the sun is vertical at that hour to the brass meridian; so it will then be in the zenith or center of the horizon. Now see what countries lie on the western edge of the horizon, for in them the sun is rising; to those on the eastern side he is setting; to those under the upper part of the meridian it is noon day; and to those under the lower part of it, it is midnight.

Thus on the 25th of April, at six o'clock in the evening, at Worcester,

The sun is rising at New Zealand; and to those who are sailing in the middle of the Great South Sea.

The sun is setting at Sweden, Hungary, Italy, Tunis, in the middle of Negroland and Guinea.

In the meridian (or noon) at the middle of Mexico, Bay of Honduras, middle of Florida, Canada, &c.

Midnight at the middle of Tartary, Bengal, India, and the seas near the Sunda isles.

PROBLEM XIII.

To find the beginning and end of twilight.

The twilight is that faint light which opens the morning by little and little in the east, before the sun rises; and gradually shuts in the evening in the west, after the sun is set. It arises from the sun's illuminating the upper part of the atmosphere, and begins always when he approaches within eighteen degrees of the eastern part of the horizon, and ends when he descends eighteen degrees below the western; when dark night commences, and continues till day breaks again.

To find the beginning of twilight, rectify the globe; turn the degree of the ecliptic, which is opposite to the sun's place, till it is elevated eighteen degrees in the quadrant of altitude above the horizon on the west, so will the index point the hour twilight begins.

This short specimen of problems by the old globes, it is presumed, will be sufficient to enable the pupil to solve any other.

PART IV.

OF THE USE OF THE CELESTIAL GLOBE.

THE celestial globe is an artificial representation of the heavens, having the fixed stars drawn upon it, in their natural order and situation; whilst it's rotation on it's axis represents the apparent diurnal motion of the sun, moon, and stars.

It is not known how early the ancients had any thing of this kind: we are not certain what the sphere of Atlas or Musæus was; perhaps Palamedes, who lived about the time of the Trojan war, had something of this kind; for of him it is said,

> To mark the signs that cloudless skies bestow,
> To tell the seasons, when to sail and plow,
> He first devised; each planet's order found,
> It's distance, period, in the blue profound.

From Pliny it would seem that Hipparchus had a celestial globe with the stars delineated upon it.

It is not to be supposed that the celestial globe is so just a representation of the heavens as the terrestrial globe is of the earth; because here the stars are drawn upon a convex surface, whereas they naturally appear in a concave one. But suppose the globe were made of glass, then to an eye placed in the center, the stars which are drawn upon it would appear in a concave surface, just as they do in the heavens.

Or if the reader was to suppose that holes were made in each star, and an eye placed in the center of the globe, it would view, through those holes, the same stars in the heavens that they represent.

As the terrestrial globe, by turning on it's axis, represents the *real diurnal motion* of the earth; so the celestial globe, by turning on it's axis, represents the *apparent diurnal motion* of the heavens.

For the sake of perspicuity, and to avoid continual references, it will be necessary to repeat here some articles which have been already mentioned.

The *ecliptic* is that graduated circle which crosses the equator in an angle of about $23\frac{1}{2}$ degrees, and the angle is called the obliquity of the ecliptic.

This circle is divided into twelve equal parts, consisting of 30 degrees each; the beginnings of them are marked with characters, representing the twelve signs.

Aries ♈, Taurus ♉, Gemini ♊, Cancer ♋, Leo ♌, Virgo ♍, Libra ♎, Scorpio ♏, Sagittarius ♐, Capricornus ♑, Aquarius ♒, Pisces, ♓.

Upon my father's globes, just under the ecliptic, the months, and days of each month, are graduated, for the readier fixing the artificial sun upon it's place in the ecliptic.

The two points where the ecliptic crosses the equinoctial, (the circle that answers to the equator on the terrestrial globe) are called the *equinoctial points;* they are at the beginnings of Aries and Libra, and are so called, because when the sun is in either of them, the day and night is every where equal.

The first points of Cancer and Capricorn are called *solstitial points;* because when the sun arrives at either of them, he seems to stand in a manner still for several days, in respect to his distance from the equinoctial; when he is in one solstitial point, he makes to us the longest day; when in the other, the longest night.

The latitude and longitude of stars are determined from the ecliptic.

The *longitude* of the stars and planets is reckoned upon the ecliptic; the numbers beginning at the first points of Aries ♈, where the ecliptic crosses the equator, and increasing according to the order of the signs.

Thus suppose the sun to be in the 10th degree of Leo, we say, his longitude, or place, is four signs, ten degrees; because he has already passed the four signs, Aries, Taurus, Gemini, Cancer, and is ten degrees in the fifth.

The *latitude* of the stars and planets is determined by their distance from the ecliptic upon a secondary or great circle passing through it's poles, and crossing it at right angles.

Twenty-four of these circular lines, which cross the ecliptic at right angles, being fifteen degrees from each other, are drawn upon the surface of our celestial globe; which being produced both ways, those on one side meet in a point on the northern polar circle, and those on the other meet in a point on the southern polar circle.

The points determined by the meeting of these circles are called the poles of the ecliptic, one north, the other south.

From these definitions it follows, that longitude and latitude, on the celestial globe, bear just the same relation to the ecliptic, as they do on the terrestrial globe to the equator.

Thus as the longitude of places on the earth is measured by degrees upon the equator, counting from the first meridian; so the longitude of the heavenly bodies is measured by degrees upon the ecliptic, counting from the first point of Aries.

And as latitude on the earth is measured by degrees upon the meridian, counting from the equator; so the latitude of the heavenly bodies is measured by degrees upon a circle of longitude, counting either north or south from the ecliptic.

The sun, therefore, *has no latitude*, being always in the ecliptic; nor do we usually speak of his longitude, but rather of his place in the ecliptic, expressing it by such a degree and minute of such a sign, as 5 degrees of Taurus, instead of 35 degrees of longitude.

The distance of any heavenly body from the equinoctial, measured upon the meridian, is called it's *declination*.

Therefore, the sun's declination, north or south, at any time, is the same as the latitude of any place to which he is then vertical, which is never more than $23\frac{1}{2}$ degrees.

Therefore all *parallels of declination* on the celestial globe are the very same as parallels of latitude on the terrestrial.

Stars may have north latitude and south declination, and vice versa.

That which is called longitude on the terrestrial globe, is called *right ascension* on the celestial; namely, the sun or star's distance from that meridian which passes through the first point of Aries, counted on the equinoctial.

Astronomers also speak of *oblique ascension* and *descension*, by which they mean the distance of that point of the equinoctial from the first point of Aries, which in an oblique sphere rises or sets, at the same time that the sun or star rises or sets.

Ascensional difference is the difference betwixt right and oblique ascension. The sun's ascensional difference turned into time, is just so much as he rises before or after six o'clock.

The celestial signs and constellations on the surface of the celestial globe, are represented by a variety of human and other figures, to which the stars that are either in or near them, are referred.

The several systems of stars, which are applied to those images, are called constellations. Twelve of these are represented on the ecliptic circle, and extend both northward and southward from it. So many of those stars as fall within the limits of 8 degrees on both sides of the ecliptic circle, together with such parts of their images as are contained within the aforesaid bounds, constitute a kind of broad hoop, belt, or girdle, which is called the *zodiac*.

The names and the respective characters of the twelve signs of the ecliptic may be learned by inspection on the surface of the broad paper circle, and the constellations from the globe itself.

The zodiac is represented by eight circles parallel to the ecliptic, on each side thereof; these circles are one degree distant from each other, so that the whole breadth of the zodiac is 16 degrees.

Amongst these parallels, the latitude of the planets is reckoned; and in their apparent motion they never exceed the limits of the zodiac.

On each side of the zodiac, as was observed, other constellations are distinguished; those on the north side are called northern, and those on the south side of it, southern constellations.

OF THE PRECESSION OF THE EQUINOXES.

All the stars which compose these constellations, are supposed to increase their longitude continually; upon which supposition, the whole starry firmament has a slow motion from west to east; insomuch that the first star in the constellation of Aries, which appeared in the vernal intersection of the equator and ecliptic in the time of Meton the Athenian, upwards of

1900 years ago, is now removed about 30 degrees from it.

This change of the stars in longitude, which has now become sufficiently apparent, is owing to a small retrograde motion of the equinoctial points, of about 50 seconds in a year, which is occasioned by the attraction of the sun and moon upon the protuberant matter about the equator. The same cause also occasions a small deviation in the parallelism of the earth's axis, by which it is continually directed towards different points in the heavens, and makes a complete revolution round the ecliptic in about 25,920 years. The former of these motions is called *the precession of the equinoxes*, the latter *the nutation of the earth's axis*. In consequence of this shifting of the equinoctial points, an alteration has taken place in the signs of the ecliptic; those stars, which in the infancy of astronomy were in Aries, being now got into Taurus, those of Taurus into Gemini, &c.; so that the stars which rose and set at any particular seasons of the year, in the times of Hesiod, Eudoxus, and Virgil, will not at present answer the descriptions given of them by those writers.

PROBLEM I.

To represent the motion of the equinoctial points backwards, or in antecedentia, upon the celestial globe, elevate the north pole so that it's axis may be perpendicular to the plane of the broad paper circle, and the equator will then be in the same plane; let these represent the ecliptic, and then the poles of the globe will also represent those of the ecliptic; the ecliptic line upon the globe will at the same time represent the equator, inclined in an angle of 23½ degrees to the broad paper circle, now called the ecliptic, and cutting it in two points, which are called the equinoctial intersections.

Now if you turn the globe slowly round upon it's axis from east to west, while it is in this position, these points of intersection will move round the same way; and the inclination of the circle, which in shewing this motion represents the equinoctial, will not be altered by such a revolution of the intersecting or equinoctial points. This motion is called the precession of the equinoxes, because it carries the equinoctial points backwards amongst the fixed stars.

The poles of the world seem to describe a circle from east to west, round the poles of the ecliptic, arising from the precession of the equinox. It is a very slow motion, for the equinoctial points take up 72 years to move one degree, and therefore they are 25,920 years in describing 360 degrees, or completing a revolution.

This motion of the poles is easily represented by the above described position of the globe, in which, if the reader remembers, the broad paper circle represents the ecliptic, and the axis of the globe being perpendicular thereto, represents the axis of the ecliptic; and the two points, where the circular lines meet, will represent the poles of the world, whence, as the globe is slowly turned from east to west, these points will revolve the same way about the poles of the globe, which are here supposed to represent the poles of the ecliptic. The axis of the world may revolve as above, although its situation, with respect to the ecliptic, be not altered; for the points here supposed to represent the poles of the world, will always keep the same distance from the broad paper circle, which represents the ecliptic in this situation of the globe.*

From the different degrees of brightness in the stars, some appear to be greater than others, or nearer to us; on our celestial globe they are distinguished into seven different magnitudes.

---

* Rutherforth's System of Nat. Philos. vol. ii. p. 730.

OF THE

USE OF THE CELESTIAL GLOBE,

IN THE SOLUTION OF

PROBLEMS RELATIVE TO THE SUN.

EVERY thing that relates to the sun is of such importance to man, that in all things he claims a natural preheminence. The sun is at once the most beautiful emblem of the Supreme Being, and, under his influence, the fostering parent of worlds; being present to them by his rays, cheering them by his countenance, cherishing them by his heat, adorning them at each returning spring with the gayest and richest attire, illuminating them with his light, and feeding the lamp of life.

To the ancients he was known under a variety of names, each characteristic of his different effects; he was their Hercules, the great deliverer, the restorer of light out of darkness, the dispenser of good, continually labouring for the happiness of a depraved race. He was the Mithra of the Persians, a word derived from love, or mercy, because the whole world is cherished by him, and feels as it were the effects of his love.

In the sacred scriptures, the original source of all emblematical writings, our Lord is called our sun, and the sun of righteousness; and as there is but one sun in the heavens, so there is but one true God, the maker and redeemer of all things, the light of the understanding, and the life of the soul.

As in scripture our God is spoken of as a shield and buckler, so the sun is characterized by this mark ☉, representing a shield or buckler, the middle point, the umbo, or boss; because it is love, or life, which alone can protect from fear and death.

His celestial rays, like those of the sun, take their circuit round the earth; there is no corner of it so remote as to be without the reach of their vivifying and penetrating power. As the material light is always ready to run it's heavenly race, and daily issues forth with renewed vigour, like an invincible champion, still fresh to labour; so likewise did our *redeeming God* rejoice to run his glorious race, he excelled in strength, and triumphed, and continues to triumph over all the powers of darkness, and is ever manifesting himself as the deliverer, the protetor, the friend, and father, of the human race.*

* Horne on the Psalms.

PLOBLEM II.

To rectify the celestial globe.

To rectify the celestial globe, is to put it in that position in which it may represent exactly the apparent motion of the heavens.

In different places, the position will vary, and that according to the different latitude of the places. Therefore, to rectify for any place, find first, by the terrestrial globe, the latitude of that place.

The latitude of the place being found in degrees, elevate the pole of the celestial globe the same number of degrees and minutes above the plane of the horizon, for this is the name given to the broad paper circle, in the use of the celestial globe.

Thus the latitude of London being $51\frac{1}{2}$ degrees, let the globe be moved till the plane of the horizon cuts the meridian in that point.

The next rectification is for the sun's place, which may be performed as directed in prob. xxix; or look for the day of the month close under the ecliptic line, against which is the sun's place, place the artificial sun over that point, then bring the sun's place to the graduated edge of the strong brazen meridian, and set the hour index to the most elevated twelve.

Thus on the 24th of May the sun is in 3½ degrees of Gemini, and is situated near the Bull's eye and the seven stars, which are not then visible, on account of his superior light. If the sun were on that day to suffer a total eclipse, these stars would then be seen shining with their accustomed brightness.

Lastly, set the meridian of the globe north and south, by the compass.

And the globe will be rectified, or put into a similar position, to the concave surface of the heavens, for the given latitude.

PROBLEM III.

To find the right ascension and declination of the sun for any day.

Bring the sun's place in the ecliptic for the given day to the meridian, and the degree of the meridian directly over it is the sun's declination for that day at noon. The point of the equinoctial cut by the meridian, when the sun's place is under it, will be the right ascension.

Thus April 19, the sun's declination is 11° 14′ north, his right ascension 27° 30′. On the 1st of December the sun's declination is 21° 54′ south, right ascension 247° 50′.

PROBLEM IV.

To find the sun's oblique ascension and descension, it's eastern and western amplitude, and time of rising and setting, on any given time, in any given place.

1. Rectify the globe for the latitude, the zenith, and the sun's place. 2. Bring the sun's place to the eastern side of the horizon; then the number of degrees intercepted between a degree of the equinoctial at the horizon and the beginning of Aries, is the sun's oblique ascension. 3. The number of degrees on the horizon intercepted between the east point and the sun's place, is the eastern or rising amplitude. 4. The hour shewn by the index is the time of sun-rising. 5. Carry the sun to the western side of the horizon, and you in the same manner obtain the oblique descension, western amplitude, and time of setting. Thus at London, May 1,

The sun's oblique ascension	18°	48'
Eastern amplitude	24	57 N
Time of rising -	4 h	40 m
Oblique descension	257°	7'
Western amplitude	26	9
Time of setting	7 h	4 m

PROBLEM V.

To find the sun's meridian altitude.

Rectify the globe for the latitude, zenith, and sun's place; and when the sun's place is in the meridian, the degrees between that point and the horizon are it's meridian altitude. Thus, on May 17, at London, the meridian altitude of the sun is 57° 55'.

PROBLEM VI.

To find the length of any day in the year, in any latitude, not exceeding 66¹ degrees.

Elevate the celestial globe to the latitude, and set the center of the artificial sun to his place upon the ecliptic line on the globe for the given day, and bring it's center to the strong brass meridian, placing the horary index to that XII which is most elevated; then turn the globe till the artificial sun cuts the eastern edge of the horizon, and the horary index will shew the time of sun-rising; turn it to the western side, and you obtain the hour of sun-setting.

The length of the day and night will be obtained by doubling the time of sun-rising and setting, as before.

PROBLEM VII.

To find the length of the longest and shortest days in any latitude that does not exceed 66½ *degrees.*

Elevate the globe according to the latitude, and place the center of the artificial sun for the longest day upon the first point of Cancer, but for the shortest day upon the first point of Capricorn; then proceed as in the last problem.

But if the place hath south latitude, the sun is in the first point of Capricorn on their longest day, and in the first point of Cancer on their shortest day.

PROBLEM VIII.

To find the latitude of a place, in which it's longest day may be of any given length between twelve and twenty-four hours.

Set the artificial sun to the first point of Cancer, bring its center to the strong brass meridian, and set the horary index to XII; turn the globe till it points to half the number of the given hours and minutes; then elevate or depress the pole till the artificial sun coincides with the horizon, and that elevation of the pole is the latitude required.

PROBLEM IX.

To find the time of the sun's rising and setting, the length of the day and night, on any place whose latitude lies between the polar circles; and also the length of the shortest day in any of those latitudes, and in what climate they are.

Rectify the globe to the latitude of the given place, and bring the artificial sun to his place in the ecliptic for the given day of the month; and then bring it's center under the strong brass meridian, and set the horary index to that XII which is most elevated.

Then bring the center of the artificial sun to the eastern part of the broad paper circle, which in this case represents the horizon, and the horary index shews the time of the sun-rising; turn the artificial sun to the western side, and the horary index will shew the time of the sun-setting.

Double the time of sun-rising is the length of the night, and the double of that of sun-setting is the length of the day.

Thus, on the 5th day of June, the sun rises at 3 h. 40 min. and sets at 8 h. 20. min.; by doubling each number it will appear, that the length of this day is 16 h. 40. min. and that of the night 7 h. 20 min.

The longest day at all places in north latitude, is when the sun is in the first point of Cancer. And,

The longest day to those in south latitude, is when the sun is in the first point of Capricorn.

Wherefore, the globe being rectified as above, and the artificial sun placed to the first point of Cancer, and brought to the eastern edge of the broad paper circle, and the horary index being set to that XII which is most elevated, on turning the globe from east to west, until the artificial sun coincides with the western edge, the number of hours counted, which are passed over by the horary index, is the length of the longest day; their complement to twenty-four hours gives the length of the shortest night.

If twelve hours be subtracted from the length of the longest day, and the remaining hours doubled, you obtain the climate mentioned by ancient historians; and if you take half the climate, and add thereto twelve hours, you obtain the length of the longest day in that climate. This holds good for every climate between the polar circles.

A climate is a space upon the surface of the earth, contained between two parallels of latitude, so far distant from each other, that the longest day in one, differs half an hour from the longest day in the other parallel.

PROBLEM X.

The latitude of a place being given in one of the polar circles, (suppose the northern) to find what number of days (of 24 hours each) the sun doth constantly shine upon the same, how long he is absent, and also the first and last day of his appearance.

Having rectified the globe according to the latitude, turn it about until some point in the first quadrant of the ecliptic (because the latitude is north) intersects the meridian in the north point of the horizon; and right against that point of the ecliptic, on the horizon, stands the day of the month when the longest day begins.

And if the globe be turned about till some point in the second quadrant of the ecliptic cuts the meridian in the same point of the horizon, it will shew the sun's place when the longest day ends, whence the day of the month may be found, as before; then the number of natural days contained between the times the longest day begins and ends, is the length of the longest day required.

Again, turn the globe about, until some

point in the third quadrant of the ecliptic cuts the meridian in the south part of the horizon; that point of the ecliptic will give the time when the longest night begins.

Lastly, turn the globe about, until some point in the fourth quadrant of the ecliptic cuts the meridian in the south point of the horizon; and that point of the ecliptic will be the place of the sun when the longest night ends.

Or, the time when the longest day or night begins being known, their end may be found by counting the number of days from that time to the succeeding solstice; then counting the same number of days from the solstitial day, will give the time when it ends.

OF THE EQUATION OF TIME.

It is not possible, in a treatise of this kind, to enter into a disquisition of the nature of time. It is sufficient to observe, that if we would with exactness estimate the quantity of any portion of infinite duration, or convey an idea of the same to others, we make use of such known measures as have been originally borrowed from the motions of the heavenly bodies. It is true, none of these motions are exactly equal and uniform, but are subject to some small irregularities, which, though of no consequence in the affairs of civil life, must be taken into the account in astronomical calculations. There are other irregularities of more importance, one of which is in the inequality of the natural day.

It is a consideration that cannot be reflected upon without surprise, that wherever we look for commensurabilities and equalities in nature, we are always disappointed. The earth is spherical, but not perfectly so; the summer is unequal, when compared with the winter; the ecliptic disagrees with the equator, and never cuts it twice in the same equinoctial point. The orbit of the earth has an eccentricity more than double in proportion to the spheroidity of it's globe; no number of the revolutions of the moon coincides with any number of the revolutions of the earth in it's orbit; no two of the planets measure one another: and thus it is wherever we turn our thoughts, so different are the views of the Creator from our narrow conception of things; where we look for commensuration, we find variety and infinity.

Thus ancient astronomers looked upon the motion of the sun to be sufficiently regular for the mensuration of time; but, by the accurate observations of later astronomers, it is found that neither the days, nor even the hours, as measured by the sun's apparent motion, are of an equal length, on two accounts.

1st, A natural or solar day of 24 hours, is that space of time the sun takes up in passing from any particular meridian to the same again; but one revolution of the earth, with respect to a fixed star, is performed in 23 hours, 56 minutes, 4 seconds; therefore the unequal progression of the earth through her elliptical orbit, (as she takes almost eight days more to run through the northern half of the ecliptic, than she does to pass through the southern) is the reason that the length of the day is not exactly equal to the time in which the earth performs it's rotation about it's axis.

2dly, From the obliquity of the ecliptic to the equator, on which last we measure time; and as equal portions of one do not correspond to equal portions of the other, the apparent motion of the sun would not be uniform; or, in other words, those points of the equator which come to the meridian, with the place of the sun on different days, would not be at equal distances from each other.

PROBLEM XI.

To illustrate, by the globe, so much of the equation of time as is in consequence of the sun's apparent motion in the ecliptic.

Bring every tenth degree of the ecliptic to the graduated side of the strong brass meridian, and you will find that each tenth degree on the equator will not come thither with it; but in the following order from ♈ to ♋, every tenth degree of the ecliptic comes sooner to the strong brass meridian than their corresponding tenths on the equator; those in the second quadrant of the ecliptic, from ♋ to ♎, come later, from ♎ to ♑ sooner, and from ♑ to Aries later, whilst those at the beginning of each quadrant come to the meridian at the same time; therefore the sun and clock would be equal at these four times, if the sun was not longer in passing through one half of the ecliptic than the other, and the two inequalities joined together, compose that difference which is called the equation of time.

These causes are independent of each other, sometimes they agree, and at other times are contrary to one another.

The inequality of the natural day is the cause that clocks or watches are sometimes before, sometimes behind the sun.

A good and well regulated clock goes uniformly on throughout the year, so as to mark the equal hours of a natural day, of a mean length; a sun-dial marks the hours of every day in such a manner, that every hour is a 24th part of the time between the noon of that day, and the noon of the day immediately following. The time measured by a clock is called equal or true time, that measured by the sun-dial apparent time.

THE USE OF THE CELESTIAL GLOBE, IN PROBLEMS RELATIVE TO THE FIXED STARS.

The use of the celestial globe is in no instance more conspicuous, than in the problems concerning the fixed stars. Among many other advantages, it will, if joined with observations on the stars themselves, render the practice and theory of other problems easy and clear to the pupil, and vastly facilitate his progress in astronomical knowledge.

The heavens are as much studded over with stars in the day, as in the night; only they are then rendered invisible to us by the brightness of the solar rays. But when this glorious luminary descends below the horizon, they begin gradually to appear; when the sun is about twelve degrees below the horizon, stars of the first magnitude become visible; when he is thirteen degrees, those of the second are seen; when fourteen degrees, those of the third magnitude appear; when fifteen degrees, those of the fourth present themselves to view; when he is descended about eighteen degrees, the stars of the fifth and sixth magnitude, and those that are still smaller, become conspicuous, and the azure arch sparkles with all it's glory.

PROBLEM XII.

To find the right ascension and declination of any given star.

Bring the given star to the meridian, and the degree under which it lies is it's declination; and the point in which the meridian intersects the equinoctial is it's right ascension. Thus the right ascension of Sirius is 99°, it's declination 16° 25′ south; the right ascension of Arcturus is 211° 32′. it's declination 20° 20′ north.

The *declination* is used to find the latitude of places; the *right ascension* is used to find the time at which a star or planet comes to the meridian; to find at any given time how long it will be before any celestial body comes to the meridian; to determine in what order those bodies pass the meridian; and to make a catalogue of the fixed stars.

PROBLEM XIII.

To find the latitude and longitude of a given star.

Bring the pole of the ecliptic to the meridian, over which fix the quadrant of altitude, and, holding the globe very steady, move the quadrant to lie over the given star, and the degree on the quadrant cut by the star, is it's latitude; the degree of the ecliptic cut at the same time by the quadrant, is the longitude of the star.

Thus the latitude of *Arcturus* is 30° 30'; it's longitude 20° 20' of Libra: the latitude of *Capella* is 22° 22' north; it's longitude 18 8' of Gemini.

The latitude and longitude of stars is used to fix precisely their place on the globe, to refer planets and comets to the stars, and, lastly, to determine whether they have any motion, whether any stars vanish, or new ones appear.

PROBLEM XIV.

The right ascension and declination of a star being given, to find it's place on the globe.

Turn the globe till the meridian cuts the equinoctial in the degree of right ascension.

178   DESCRIPTION AND USE

Thus for example, suppose the right ascension of Aldebaran to be 65° 30′, and it's declination to be 16° north, then turn the globe about till the meridian cuts the equinoctial in 65° 30′, and under the 16° of the meridian, on the northern part, you will observe the star Aldebaran, or the bull's eye.

PROBLEM XV.

To find at what hour any known star passes the meridian, at any given day.

Find the sun's place for that day in the ecliptic, and bring it to the strong brass meridian, set the horary index to XII o'clock, then turn the globe till the star comes to the meridian, and the index will mark the time. Thus on the 15th of August, Lyra comes to the meridian at 45 min. past VIII in the evening. On the 14th of September the brightest of the Pleiades will be on the meridian at IV in the morning.

This problem is useful for directing when to look for any star on the meridian, in order to find the latitude of a place, to adjust a clock, &c.

PROBLEM XVI.

To find on what day a given star will come to the meridian, at any given hour.

Bring the given star to the meridian, and set the index to the proposed hour; then turn the globe till the index points to XII at noon, and observe the degree of the ecliptic then at the meridian; this is the sun's place, the day answering to which may be found on the calendar of the broad paper circle.

By knowing whether the hour be in the morning or afternoon, it will be easy to perceive which way to turn the globe, that the proper XII may be pointed to; the globe must be turned towards the west, if the given hour be in the morning, towards the east if it be afternoon.

Thus Arcturus will be on the meridian at III in the morning on March the 5th, and Cor Leonis at VIII in the evening on April the 21st.

PROBLEM XVII.

To represent the face of the heavens on the globe for a given hour on any day of the year, and learn to distinguish the visible fixed stars.

Rectify the globe to the given latitude of the place and day of the month, setting it due north and south by the needle; then turn the globe on it's axis till the index points to the given hour of the night; then all the upper hemisphere of the globe will represent the visible face of the heavens for that time, by which it will be easily seen what constellations and stars of note are then above our horizon, and what position they have with respect to the points of the compass. In this case, supposing the eye was placed in the center of the globe, and holes were pierced through the centers of the stars on it's surface, the eye would perceive through those holes the various corresponding stars in the firmament; and hence it would be easy to know the various constellations at sight, and to be able to call all the stars by their names.

Observe some star that you know, as one of the pointers in the Great Bear, or Sirius; find the same on the globe, and take notice of the position of the contiguous stars in the same or an adjoining constellation; direct your sight to the heavens, and you will see those stars in the same situation. Thus you may proceed from one constellation to another, till you are acquainted with most of the principal stars.

" For example: the situation of the stars at London on the 9th of February, at 2 min. past IX in the evening, is as follows.

" Sirius, or the Dog-star, is on the meridian,

it's altitude 22°: Procyon, or the little Dog-star, 16° towards the east, it's altitude 43½: about 24° above this last, and something more towards the east, are the twins, Castor and Pollux: S.6 5° E. and 35° in height, is the bright star Regulus, or Cor Leonis: exactly in the east and 22° high, is the star Deneb Alased in the Lion's tail: 30° from the east towards the north Arcturus is about 3 above the horizon: directly over Arcturus, and 31° above the horizon, is Cor Caroli: in the north-east are the stars in the extremity of the Great Bear's tail, Aleath the first star in the tail, and Dubhe the northernmost pointer in the same constellation; the altitude of the first of these is 30½, that of the second 41°, and that of the third 56°.

"Reckoning westward, we see the beautiful constellation Orion; the middle star of the three in his belt, is S. 20° W. it's altitude 35°: nine degrees below the belt, and a little more to the west, is Rigel the bright star in his heel: above his belt in a strait line drawn from Rigel between the middle and most northward in his belt, and 9° above it, is the bright star in his shoulder: S. 49° W. and 45½ above the horizon, is Aldebaran the southern eye of the Bull: a little to the west of Aldebaran, are the Hyades: the same altitude, and about S. 70° W, are the Pleiades: in the W. by S. point is Capella in Auriga, it's altitude 73°: in the north-west, and about 42° high, is the constellation Cassiopeia: and almost in the north, near the horizon, is the constellation Cygnus."*

PROBLEM XVIII.

To trace the circles of the sphere in the starry firmament.

I shall solve this problem for the time of the autumnal equinox; because that intersection of the equator and ecliptic will be directly under the depressed part of the meridian about midnight; and then the opposite intersection will be elevated above the horizon; and also because our first meridian upon the terrestrial globe passing through London, and the first point of Aries, when both globes are rectified to the latitude of London, and to the sun's place, and the first point of Aries is brought under the graduated side of each of their meridians, we shall have the corresponding face of the heavens and the earth represented, as they are with respect to each other at that time, and the principal circles of each sphere will correspond with each other.

The horizon is then distinguished, if we begin from the north, and count westward, by the following constellations; the hounds and waist of Bootes, the northern crown, the head of

* Bransby's Use of the Globes.

Hercules, the shoulders of Serpentarius, and Sobieski's shield; it passes a little below the feet of Antinous, and through those of Capricorn, through the Sculptor's frame, Eridanus, the star Rigel in Orion's foot, the head of Monoceros, the Crab, the head of the Little Lion, and lower part of the Great bear.

The meridian is then represented by the equinoctial colure, which passes through the star marked δ in the tail of the Little Bear, under the north pole, the pole star, one of the stars in the back of Cassiopeia's chair marked β, the head of Andromeda, the bright star in the wing of Pegasus marked γ, and the extremity of the tail of the whale.

That part of the equator which is then above the horizon, is distinguished on the western side by the northern part of Sobieski's shield, the shoulder of Antinous, the head and vessel of Aquarius, the belly of the western fish in Pisces; it passes through the head of the Whale, and a bright star marked δ in the corner of his mouth, and thence through the star marked δ in the belt of Orion, at that time near the eastern side of the horizon.

That half of the ecliptic which is then above the horizon, if we begin from the western side, presents to our view Capricornus, Aquarius, Pisces, Aries, Taurus, Gemini, and a part of the constellation Cancer.

The solstitial colure, from the western side, passes through Cerberus, and the hand of Hercules, thence by the western side of the constellation Lyra, and through the Dragon's head and body, through the pole point under the polar star, to the east of Auriga, through the star marked η in the foot of Castor, and through the hand and elbow of Orion.

The northern polar circle, from that part of the meridian under the elevated pole, advancing towards the west, passes through the shoulder of the Great Bear, thence a little to the north of the star marked α in the Dragon's tail, the great knot of the dragon, the middle of the body of Cepheus, the northern part of Cassiopeia, and base of her throne, through Camelopardalus, and the head of the Great Bear.

The tropic of Cancer, from the western edge of the horizon, passes under the arm of Hercules, under the Vulture, through the Goose and Fox, which is under the beak and wing of the Swan, under the star called Sheat, marked β in Pegasus, under the head of Andromeda, and through the star marked Φ in the fish of the constellation Pisces, above the bright star in the head of the Ram marked α, through the Pleiades, between the horns of the Bull, and through a group of stars at the foot of Castor, thence above a star marked δ, between Castor and Pollux, and so through a part of the constellation
370

Cancer, where it disappears by passing under the eastern part of the horizon.

The tropic of Capricorn, from the western side of the horizon, passes through the belly, and under the tail of Capricorn, thence under Aquarius, through a star in Eridanus marked c, thence under the belly of the Whale, through the base of the chemical Furnace, whence it goes under the Hare at the feet of Orion, being there depressed under the horizon.

The southern polar circle is invisible to the inhabitants of London, by being under our horizon.

Arctic and antarctic circles, or circles of perpetual apparition and occultation.

The largest parallel of latitude on the terrestrial globe, as well as the largest circle of declination on the celestial, that appears entire above the horizon of any place in north latitude, was called by the ancients the arctic circle, or circle of perpetual apparition.

Between the arctic circle and the north pole in the celestial sphere, are contained all those stars which never set at that place, and seem to us, by the rotative motion of the earth, to be perpetually carried round above our horizon, the circles parallel to the equator.

The largest parallel of latitude on the terrestrial, and the largest parallel of declination on the celestial globe, which is entirely hid below the horizon of any place, was by the ancients called the antarctic circle, or circle of perpetual occultation.

This circle includes all the stars which never rise in that place to an inhabitant of the northern hemisphere, but are perpetually below the horizon.

All arctic circles touch their horizons in the north point, and all antarctic circles touch their horizons in the south point; which point, in the terrestrial and celestial spheres, is the intersection of the meridian and horizon.

If the elevation of the pole be 45 degrees, the most elevated part either of the arctic or antarctic circle will be in the zenith of the place.

If the pole's elevation be less than 45 degrees, the zenith point of those places will fall without it's arctic or antarctic circle; if greater, it will fall within.

Therefore, the nearer any place is to the equator, the less will it's arctic and antarctic circles be; and on the contrary, the farther any place is from the equator, the greater they are. So that,

At the poles, the equator may be considered as both an arctic and antarctic circle, because it's plane is coincident with that of the horizon.

But at the equator (that is, in a right sphere) there is neither arctic nor antarctic circle.

They who live under the northern polar circle, have the tropic of Cancer for their arctic, and that of Capricorn for their antarctic circle.

And they who live on either tropic, have one of the polar circles for their arctic, and the other for their antarctic circle.

Hence, whether these circles fall within or without the tropics, their distance from the zenith of any place is ever equal to the difference between the pole's elevation, and that of the equator, above the horizon of that place.

From what has been said, it is plain there may be as many arctic and antarctic circles, as there are individual points upon any one meridian, between the north and south poles of the earth.

Many authors have mistaken these mutable circles, and have given their names to the immutable polar circles, which last are arctic and antarctic circles, in one particular case only, as has been shewn.

PROBLEM XIX.

To find the circle, or parallel of perpetual apparition, or occulation of a fixed star, in a given latitude.

By rectifying the globe to the latitude of the place, and turning it round on it's axis, it will be immediately evident, that the circle of perpetual apparition is that parallel of declination which is equal to the complement of the given latitude northward; and for the perpetual occultation, it is the same parallel southward; that is to say, in other words, all those stars, whose declinations exceed the co-latitude, will always be visible, or above the horizon; and all those in the opposite hemisphere, whose declination exceeds the co-latitude, never rise above the horizon.

For instance; in the latitude of London 51 deg. 30 min. whose co-latitude is 38 deg. 30 min. gives the parallels desired; for all those stars which are within the circle, towards the north pole, never descend below our horizon; and all those stars which are within the same circle, about the south pole, can never be seen in the latitude of London, as they never ascend above it's horizon.

OF PROBLEMS RELATING TO THE AZIMUTH, &c.
OF THE SUN AND STARS.

PROBLEM XX.

The latitude of the place and the sun's place being given, to find the sun's amplitude.

That degree from east or west in the horizon, wherein any object rises or sets, is called the amplitude.

Rectify the globe, and bring the sun's place to the eastern side of the meridian, and the arch of the horizon intercepted between that point and the eastern point, will be the sun's amplitude at rising.

If the same point be brought to the western side of the horizon, the arch of the horizon intercepted between that point and the western point, will be the sun's amplitude at setting.

Thus on the 24th of May the sun rises at four, with 36 degrees of eastern amplitude, that is, 36 degrees from the east towards the north, and sets at eight, with 36 degrees of western amplitude.

The amplitude of the sun at rising and setting increases with the latitude of the place: and in very high northern latitudes, the sun scarce sets before he rises again. Homer had heard something of this, though it is not true of the Læstrygones, to whom he applies it.

> Six days and nights a doubtful course we steer;
> The next, proud Lamos' lofty towers appear,
> And Læstygonia's gates arise distinct in air.
> The shepherd quitting here at night the plain,
> Calls, to succeed his cares, the watchful swain.
> But he that scorns the chains of sleep to wear,
> And adds the herdsman's to the shepherd's care,
> So near the pastures, and so short the way,
> His double toils may claim a double pay,
> And join the labours of the nigiht and day.

PROBLEM XXI.

To find the sun's altitude at any given time of the day.

Set the center of the artificial sun to his place in the ecliptic upon the globe, and rectify it to the latitude and zenith; bring the center of the artificial sun under the strong brass meridian, and set the hour index to that XII which is most elevated; turn the globe to the given hour, and move the graduated edge of the quadrant to the center of the artificial sun; and that degree on the quadrant, which is cut by the sun's center, is the sun's height at that time.

The artificial sun being brought under the strong brass meridian, and the quadrant laid upon it's center, will *shew it's meridian, or greatest altitude,* for that day.

If the sun be in the equator, his greatest or meridian altitude is equal to the elevation of the equator, which is always equal to the co-latitude of the place.

Thus on the 24th of May, at nine o'clock, the sun has 44 deg. altitude, and at six in the afternoon 20 deg.

OF THE AZIMUTHAL OR VERTICAL CIRCLES.

The vertical point, that is, the uppermost point of the celestial globe, represents a point in the heavens, directly over our heads, which is called our zenith.

From this point circular lines may be conceived crossing the horizon at right angles.

These are called *azimuth*, or *vertical circles*. That one which crosses the horizon at 10 deg. distance from the meridian on either side, is called an azimuth circle of 10 deg.; that which crosses at 20, is called an azimuth of 20 deg.

The azimuth of 90 deg. is called the *prime vertical:* it crosses the horizon at the eastern and western points.

Any *azimuth circle* may be represented by the graduated edge of the brass quadrant of altitude, when the center upon which it turns is screwed to that point of the strong brass meridian which answers to the latitude of the place, and the place is brought into the zenith.

If the said graduated edge should lie over the sun's center or place, at any given time, it will represent the sun's azimuth at that time.

If the graduated edge be fixed at any point, so as to represent any particular azimuth, and the sun's place be brought there, the horary index will shew at what time of that day the sun will be in that particular azimuth.

Here it may be observed, that the *amplitude* and azimuth are much the same.

The amplitude shewing the bearing of any object *when it rises or sets*, from the *east* and *west* points of the horizon.

The azimuth the bearing of any object when it is *above the horizon*, either from the *north* or *south* points thereof. These descriptions and illustrations being understood, we may proceed to

PROBLEM XXII.

To find at what time the sun is due east, the day and the latitude being given.

Rectify the globe; then if the latitude and declination are of one kind, bring the quadrant of altitude to the eastern point of the horizon, and the sun's place to the edge of the quadrant, and the index will shew the hour.

If the latitude and declination are of different kinds, bring the quadrant to the western point of the horizon, and the point in the ecliptic opposite to the sun's place to the edge of the quadrant, and then the index will shew the hour.

You will easily comprehend the reason of the foregoing distinction, because when the sun is in the equinoctial, it rises due east; but when it is in that part of the ecliptic which is towards the elevated pole, it rises before it is in the eastern vertical circle, and is therefore at that time *above* the horizon: whereas when it is in the other part of the ecliptic, it passes the eastern prime vertical before it rises, that is *below* the horizon; whence it is evident, that the opposite point of the ecliptic must then be in the west, and above the horizon. The sun is due east at London at 7 h. 6 min. on the 18th of May. The second of August, at Cape Horn, the sun is due east at 5 h. 10 min.

PROBLEM XXIII.

To find the rising, setting, and culminating of a star, it's continuance above the horizon, and it's oblique ascension and descension, and also it's eastern and western amplitude, for any given day and place.

1. Rectify the globe to the latitude and zenith, bring the sun's place for the day to the meridian, and set the hour index to XII. 2. Bring the star to the eastern side of the horizon, and it's eastern amplitude, oblique ascension, and time of rising, will be found as taught of the sun. 3. Carry the star to the western side of the horizon; and in the same manner it's western amplitude, oblique descension, and time of setting, will be found. 4. The time of rising, subtracted from that of setting, leaves the continuance of the star above the horizon. 5. This remainder, subtracted from 24 hours, gives the time of it's continuance below the horizon. 6. The hour to which the index points, when the star comes to the meridian, is the time of it's culminating or being on the meridian.

Let the given day be March 14, the place

London, the star Sirius; by working the problem, you will find

It rises at - - 2 h. 24 min. afternoon.
Culminates - - 6 57
Sets at - - 11 50
Is above the horizon 9 6

It's oblique ascension and descension are 120° 47', and 77° 15'; it's amplitude 27°, southward.

PROBLEM XXIV.

The latitude, the altitude of the SUN *by day, or of a* STAR *by night, being given, to find the hour of the day, and the sun's or star's azimuth.*

Rectify the globe for the latitude, the zenith, and the sun's place, turn the globe and the quadrant of altitude, so that the sun's place, or the given star, may cut the given degree of altitude, the index will shew the hour, and the quadrant will be the azimuth in the horizon.

Thus on the 21st of August, at London, when the sun's altitude is 36° in the forenoon, the hour is IX, and the azimuth 58° from the south.

At Boston, December 8th, when Rigel had 15 of altitude, the hour was VIII, the azimuth S. E. by E. 7°.

PROBLEM XXV.

The latitude and hour of the day being given, to find the altitude and azimuth of the sun, or of a star.

Rectify the globe for the latitude, the zenith, and the sun's place, then the number of degrees contained betwixt the sun's place and the vertex is the sun's meridional zenith distance; the complement of which to 90 deg. is the sun's meridian altitude. If you turn the globe about until the index points to any other given hour, then bringing the quadrant of altitude to cut the sun's place, you will have the sun's altitude at that hour; and where the quadrant cuts the horizon, is the sun's azimuth at the same time. Thus May the 1st, at London, the sun's meridian altitude will be $53\frac{1}{2}$ deg.; and at 10 o'clock in the morning, the sun's altitude will be 46 deg. and his azimuth about 44 deg. from the south part of the meridian. On the 2d of December, at Rome, at five in the morning, the altitude of Capella is 41 deg. 58 min. its azimuth 60 deg. 50 min. from N. to W.

PROBLEM XXVI.

The latitude of the place, and the day of the month being given, to find the depression of the sun below the horizon, and the azimuth, at any hour of the night.

Having rectified the globe for the latitude, the zenith, and the sun's place, take a point in the ecliptic exactly opposite to the sun's place, and find the sun's altitude and azimuth, as by the last problem, and these will be the depression and the altitude required.

Thus if the time given be the 1st of November, at 10 o'clock at night, the depression and azimuth will be the same as was found in the last problem.

PROBLEM XXVII.

The latitude, the sun's place, and his azimuth being given, to find his altitude, and the hour.

Rectify the globe for the latitude, the zenith, and the sun's place; then put the quadrant of altitude to the sun's azimuth in the horizon, and turn the globe till the sun's place meets the edge of the quadrant; then the said edge will shew the altitude, and the index point to the hour.

Thus, May 21st, at London, when the sun

is due east, his altitude will be about 24 deg. and the hour about VII in the morning; and when his azimuth is 60 degrees south-westerly, the altitude will be about 44½ degrees, and the hour II¾ in the afternoon.

Thus the latitude and the day being known, and having besides either the altitude, the azimuth, or the hour, the other two may be easily found.

PROBLEM XXVIII.

The latitude of the place, and the azimuth of the sun or of a star being given, to find the hour of the day or night.

Rectify the globe for the latitude and sun's place, and bring the quadrant of altitude to the given azimuth in the horizon; turn the globe till the sun or star comes to the quadrant, and the index will shew the time. November 5, at Gibraltar, given the sun's azimuth 50 degrees from the south towards the east, the time you will find to be half past VIII in the morning. Given the azimuth of Vega at London, 57 deg. from the north towards the east, February the 8th, the time you will find twenty minutes past II in the morning.

But as it may possibly happen that we may see a star, and would be glad to know what star it is, or whether it may not be a new star, or a comet; how that may be discovered, will be seen under the following

PROBLEM XXIX.

The latitude of the place, the sun's place, the hour of the night, and the altitude and azimuth of any star being given, to find the star.

Rectifying the globe for the latitude of the place, and the sun's place; fix the quadrant of altitude in the zenith, and turn the globe till the hour index points to the given hour, and set the quadrant of altitude to the given azimuth; then the star that cuts the quadrant in the given altitude, will be the star sought.

Though two stars, that have different right ascensions, will not come to the meridian at the same time, yet it is possible that in a certain latitude they may come to the same vertical circle at the same time; and that consideration gives the following

PROBLEM XXX.

The latitude of the place, the sun's place, and two stars, that have the same azimuth, being given, to find the hour of the night.

Rectify the globe for the latitude, the zenith, and the sun's place; then turn the globe, and also the quadrant about, till both the stars coincide with it's edge; the hour index will shew the hour of the night, and the place where the quadrant cuts the horizon will be the common azimuth of both stars.

On the 15th of March, at London, the star Betelgeule, in the shoulder of Orion, and Regel, in the heel of Orion, were observed to have the same azimuth; on working the problem, you will find the time to be 8 hours 47 minutes.

What hath been observed above, of two stars that have the same azimuth, will hold good likewise of two stars that have the same altitude; from whence we have the following

PROBLEM XXXI.

The latitude of the place, the sun's place, and two stars, that have the same altitude, being given, to find the hour of the night.

Rectify the globe for the latitude of the place, the zenith, and the sun's place; turn the globe, so that the same degree on the quadrant shall cut both the stars, then the hour index will shew the hour of the night.

In the former propositions, the latitude of the place is supposed to be given, or known; but as it is frequently necessary to find the latitude of the place, especially at sea, how this may be found, in a rude manner at least, having the time given by a good clock, or watch, will be seen in the following.

PROBLEM XXXII.

The suns's place, the hour of the night, and two stars, that have the same azimuth, or altitude, being given, to find the latitude of the place.

Rectify the globe for the sun's place, and turn it till the index points to the given hour of the night; keep the globe from turning, and move it up and down in the notches, till the two given stars have the same azimuth, or altitude; then the brass meridian will shew the height of the pole, and consequently the latitude of the place.

PROBLEM XXXII.

Two stars being given, one on the meridian, and the other on the east and west part of the horizon, to find the latitude of the place.

Bring the star observed on the meridian to the meridian of the globe; then keeping the globe from turning round it's axis, slide the meridian up or down in the notches, till the other star is brought to the east or west part of the horizon, and that elevation of the pole will be the latitude of the place sought.

OBSERVATION.

From what hath been said, it appears, that of these five things, 1. the latitude of the place; 2. the sun's place in the ecliptic; 3. the hour of the night; 4. the common azimuth of two known fixed stars; 5. the equal altitude of two known fixed stars; any *three* of them being given, the remaining *two* will easily be found.

There are three sorts of risings and settings of the fixed stars, taken notice of by ancient authors, and commonly called *poetial risings* and *settings*, because mostly taken notice of by the poets.

These are the *cosmical*, *achronical*, and *heliacal*.*

They are to be found in most authors that treat on the doctrine of the sphere, and are now chiefly useful in comparing and understanding passages in the ancient writers; such are Hesiod, Virgil, Columella, Ovid, Pliny, &c. How they are to be found by calculation, may be seen in Petavius's Uranologion, and Dr. Gregory's Astronomy.

DEFINITION.

When a star rises or sets at sun-rising, it is said to rise or set COSMICALLY.

From whence we shall have the following

---

* Costard's History of Astronomy.

PROBLEM XXXIV.

The latitude of the place being given, to find, by the globe, the time of the year when a given star rises or sets cosmically.

Let the given place be Rome, whose latitude is 42 deg. 8 min. north; and let the given star be the Lucida Pleiadum. Rectify the globe for the latitude of the place; bring the star to the edge of the eastern horizon, and mark the point of the ecliptic rising along with it; that will be found to be Taurus, 18 deg. opposite to which, on the horizon, will be found May the 8th. The Lucida Pleiadum, therefore, rises cosmically May the 8th.

If the globe continues rectified as before, and the Lucida Pleiadum be brought to the edge of the western horizon, the point of the ecliptic, which is the sun's place, then rising on the eastern side of the horizon, will be Scorpio, 29 deg. opposite to which, on the horizon, will be found November the 21st. The Lucida Pleiadum, therefore, sets cosmically November the 21st.

In the same manner, in the latitude of London, Sirius will be found to rise cosmically August the 10th, and to set cosmically November the 10th.

It is of the cosmical setting of the Pleiades,

that Virgil is to be understood in this line,
*Ante tibi Eoæ Atlantides abscondantur,**
and not of their *setting in the east,* as some have imagined, where stars rise, but never set.

DEFINITION.

When a star rises or sets at sun-setting, it is said to rise or set ACHRONICALLY.

Hence, likewise, we have the following

PROBLEM. XXXV.

The latitude of the place being given, to find the time of the year when a given star will rise or set achronically.

Let the given place be Athens, whose latitude is 37 deg. north, and let the given star be Arcturus.

Rectify the globe for the latitude of the place, and bringing Arcturus to the eastern side of the horizon, mark the point of the ecliptic then setting on the western side; that will be found Aries, 12 deg. opposite to which, on the horizon, will be found April the 2d. Therefore Arcturus rises at Athens achronically April the 2d.

It is of this rising of Arcturus that Hesiod speaks in his Opera and Dies.†

> When from the solstice sixty wint'ry days
> Their turns have finish'd, mark, with glitt'ring rays,
> From ocean's sacred flood, *Arcturus* rise,
> Then first to gild the dusky evening skies.

* Georg. l. 1. v. 221.   † Lib. ii. ver. 285.

If the globe continues rectified to the latitude of the place, as before, and Arcturus be brought to the western side of the horizon, the point of the ecliptic setting along with it will be Sagittary, 7 deg. opposite to which, on the horizon, will be found November the 29th. At Athens, therefore, Arcturus sets achronically November the 29th.

In the same manner Aldebaran, or the Bull's eye, will be found to rise achronically May the 22d, and to set achronically December the 19th.

DEFINITION.

When a star first becomes visible in a morning, after it hath been so near the sun as to be hid by the splendor of his rays, it is said to rise HELIACALLY.

But for this there is required some certain depression of the sun below the horizon, more or less according to the magnitude of the star. A star of the first magnitude is commonly supposed to require that the sun be depressed 12 deg. perpendicularly below the horizon.

This being premised, we have the following

PROBLEM XXXVI.

The latitude of the place being given, to find the time of the year when a given star will rise heliacally.

Let the given place be Rome, whose latitude is 42 deg. north, and let the given star be the bright star in the Bull's horn.

Rectify the globe for the latitude of the place, screw on the brass quadrant of altitude in it's zenith, and turn it to the western side of the horizon. Bring the star to the eastern side of the horizon, and mark what degree of the ecliptic is cut by 12 deg. marked on the quadrant of altitude; that will be found to be Capricorn, 3 deg. the point opposite to which is Cancer, 3 deg. and opposite to this will be found on the horizon, June 25th. The bright star, therefore, in the Bull's horn, in the latitude of Rome, rises heliacally June the 25th.

These kinds of risings and settings are not only mentioned by the poets, but likewise by the ancient physicians and historians.

Thus Hippocrates, in his book De Æ re, says, " One ought to observe the heliacal risings and settings of the stars, especially the *Dog-star*, and *Arcturus*; likewise the *cosmical* setting of the *Pleiades*."

And Polybius, speaking of the loss of the

Roman fleet, in the first Punic war, says, " It was not so much owing to fortune, as to the obstinacy of the consuls, in not hearkening to their pilots, who dissuaded them from putting to sea, at that season of the year, which was between the rising of *Orion* and the *Dog-star ;* it being always dangerous, and subject to storms."*

DEFINITION.

When a star is first immersed in the evening, or hid by the sun's rays, it is said to set HELIA- CALLY.

And this again is said to be, when a star of the first magnitude comes within twelve degrees of the sun, reckoned in the perpendicular.

Hence again we have the following

PROBLEM XXXVII.

The latitude of the place being given, to find the time of the year when a given star sets helia- cally.

Let the given place be Rome, in latitude 42 deg. north, and let the given star be the bright star in the Bull's horn. Rectify the globe for the latitude of the place, and bring the star

* Lib. i. p. 53.

to the edge of the western horizon; turn the quadrant of altitude, till 12 deg. cut the ecliptic on the eastern side of the meridian. This will be found to be 7 deg. of Sagittary, the point opposite to which, in the ecliptic, is 7 deg. of Gemini; and opposite to that, on the horizon, is May the 28th, the time of the year when that sets heliacally in the latitude of Rome.

OF THE CORRESPONDENCE OF THE CELESTIAL AND TERRESTRIAL SPHERES.

That the reader may thoroughly understand what is meant by the correspondence between the two spheres, let him imagine the celestial globe to be delineated upon glass, or any other transparent matter, which shall invest or surround the terrestrial globe, but in such a manner, that either may be turned about upon the poles of the globe, while the other remains fixed; and suppose the first point of Aries, on the investing globe, to be placed on the first point of Aries on the terrestrial globe, (which point is in the meridian of London) then every star in the celestial sphere will be directly over those places to which it is a correspondent. Each star will then have the degree of it's right ascension directly upon the corresponding degree of terrestrial longitude; their declination will also be the same with the latitude of the places to which they answer; or, in other words, when the declination of a star is equal to the latitude of a place, such star, within the space of 24 hours, will pass vertically over that place and all others that have the same latitude.

If we conceive the celestial investing globe to to be fixed, and the terrestrial globe to be gradually turned from west to east, it is clear, that as the meridian of London passes from one degree to another under the investing sphere, every star in the celestial sphere becomes correspondent to another place upon the earth, and so on, until the earth has completed one diurnal revolution; or till all the stars, by their apparent daily motion, have passed over every meridian of the terrestrial globe. From this view of the subject, an amazing variety, uniting in wonderful and astonishing harmony, presents itself to the attentive reader; and future ages will find it difficult to investigate the reasons that should induce the present race of astronomers to neglect a subject so highly interesting to science, even in a practical view, but which in theory would lead them into more sublime speculations, than any that ever yet presented themselves to their minds.

A GENERAL DESCRIPTION OF THE PASSAGE OF THE STAR MARKED γ IN THE HEAD OF THE CONSTELLATION DRACO, OVER THE PARALLEL OF LONDON.

The star γ, in the head of the constellation Draco, having 51 deg. 32 min. north declination, equal to the latitude of London, is the correspondent star thereto. To find the places which it passes over, bring London to the graduated side of the brass meridian, and you will find that the degree of the meridian over London, and the representative of the star, passes over from London, the road to Bristol, crosses the Severn, the Bristol channel, the counties of Cork and Kerry in Ireland, the north part of the Atlantic ocean, the streights of Belleisle, New Britain, the north part of the province of Canada, New South Wales, the southern part of Kamschatka, thence over different Tartarian nations, several provinces of Russia, over Poland, part of Germany, the southern part of the United Provinces, when, crossing the sea, it arrives again at the meridian of London.

When the said star, or any other star, is on the meridian of London, or any other meridian, all other stars, according to their declination and right ascension, and difference of right ascension, (which answers to terrestrial latitude, longitude, and difference of longitude) will at the same time be on such meridians, and vertical to such places as correspond in latitude, longitude, and difference of longitude, with the declination, &c. of the respective stars.*

From the stars, therefore, thus considered, we attain a copious field of geographical knowledge, and may gain a clear idea of the proportionable distances and real bearings, of remote empires, kingdoms, and provinces, from our own zenith, at the same instant of time; which may be found in the same manner as we found the place to which the sun was vertical at any proposed time.

Many instances of this mode of attaining geographical knowledge, may be found in my father's treatise on the globes.

OF THE USE OF THE CELESTIAL GLOBE, IN PROBLEMS RELATIVE TO THE PLANETS.

The situation of the fixed stars being always the same with respect to one another, they have their proper places assigned to them on the globe.

But to the planets no certain place can be assigned, their situation always varying.

* Fairman's Geography.

That space in the heavens, within the compass of which the planets appear, is called the zodiac.

The latitude of the planets scarce ever exceeding 8 degrees, the zodiac is said to reach about 8 degrees on each side the ecliptic.

Upon the celestial globe, on each side of the ecliptic, are drawn eight parallel circles, at the distance of one degree from each other, including a space of 16 degrees; these are crossed at right angles, with segments of great circles at every 5th degree of the ecliptic; by these, the place of a planet on the globe, on any given day, may be ascertained with accuracy.

PROBLEM XXXVIII.

To find the place of any planet upon the globe, and by that means to find it's place in the heavens: also, to find at what hour any planet will rise or set, or be on the meridian, on any day in the year.

Rectify the globe to the latitude and sun's place, then place the planet's longitude and latitude in an ephemeris, and set the graduated edge of the moveable meridian to the given longitude in the ecliptic, and counting so many degrees amongst the parallels in the zodiac, either above or below the ecliptic, as her latitude is north or south; and set the center of the artificial sun to that point, and the centre will represent the place of the planet for that time.

Or fix the quadrant of altitude over the pole of the ecliptic, and holding the globe fast, bring the edge of the quadrant to cut the given degree of longitude on the ecliptic; then seek the given latitude on the quadrant, and the place under it is the point sought.

While the globe moves about it's axis, this point moving along with it will represent the planet's motion in the heavens. If the planet be brought to the eastern side of the horizon, the horary index will shew the time of it's rising. If the artificial sun is above the horizon, the planet will not be visible: when the planet is under the strong brazen meridian, the hour index shews the time it will be on that circle in the heavens: when it is at the western edge, the time of it's setting will be obtained.

PROBLEM XXXIX.

To find directly the planets which are above the horizon at sun-set, upon any given day and latitude.

Find the sun's place for the given day, bring it to the meridian, set the hour index to XII, and elevate the pole for the given latitude: then bring the place of the sun to the western semicircle of the horizon, and observe what signs are in that part of the ecliptic above the horizon, then cast your eye upon the ephemeris for that month, and you will at once see what planets possess any of those elevated signs; for such will be visible, and fit for observation on the night of that day.

PROBLEM XL.

To find the right ascension, declination, amplitude, azimuth, altitude, hour of the night, &c. of any given planet, for a day of a month and latitude given.

Rectify the globe for the given latitude and day of the month; then find the planet's place, as before directed, and then the right ascension, declination, amplitude, azimuth, altitude, hour, &c. are all found, as directed in the problems for the sun; there being no difference in the process, no repetition can be necessary.

OF THE USE OF THE CELESTIAL GLOBE, IN PROBLEMS RELATIVE TO THE MOON.

From the sun and planets we now proceed to those problems that concern the moon, the brilliant satellite of our earth, which every month enriches it with it's presence; by the mildness of it's light softening the darkness of night; by it's influence affecting the tide; and by the variety of it's aspects, offering to our view some very remarkable phenomena.

> " Soon as the ev'ning shades prevail,
> The moon takes up the wond'rous tale :
> And nightly to the list'ning earth,
> Repeats the story of her birth :
> Whilst all the stars that round her burn,
> And all the planets in their turn,
> Confirm the tidings as they roll,
> And spread the truth from pole to pole."

As the orbit of the moon is constantly varying in its position, and the place of the node always changing, as her motion is even variable in every part of her orbit, the solutions of the problems which relate to her, are not altogether so simple as those which concern the sun.

The moon increases her longitude in the ecliptic every day, about 13 degrees, 10 minutes, by which means she crosses the meridian of any place about 50 minutes later than she did the preceding day.

Thus if on any day at noon her place (longitude) be in the 12th degree of Taurus, it will be 13 deg. 10 min. more, or 25 deg. 10 min. in Taurus on the succeeding noon.

It is new moon when the sun and moon

have the same longitude, or are in or near the same point of the ecliptic.

When they have opposite longitudes, or are in opposite points of the ecliptic, it is full moon.

To ascertain the moon's place with accuracy, we must recur to an ephemeris; but as even in most ephemerides the moon's place is only shewn at the beginning of each day, or XII o'clock at noon, it becomes necessary to supply by a table this deficiency, and assign thereby her place for any intermediate time.

In the nautical ephemeris, published under the authority of the Board of Longitude, we have the moon's place for noon and midnight, with rules for accurately obtaining any intermediate time; but as this ephemeris may not always be at hand, we shall insert, from Mr. Martin's treatise on the globes, a table for finding the hourly motion of the moon. In order, however, to use this table, it will be necessary first *to find the quantity of the moon's diurnal motion in the ecliptic*, for any given day; for the quantity of the moon's diurnal motion varies from about 11 deg. 46 min. the least, to 15 deg. 16 min. when greatest.

The following tables are calculated from the least of 11 deg. 46 min. to the greatest of 15 deg. 16 min. every column increasing 10 minutes; upon the top of the column is the quantity of the diurnal motion, and on the side of the table are the 24 hours, by which means it will be easy to find what part of the diurnal motion of the moon answers to any given number of hours.

Thus suppose the diurnal motion to be 12° 32′, look on the top column for the number nearest to it, which you will find to be 12° 36′, in the sixth column; and under it, against 9 hours, you will find 4 deg. 43 min. which is her motion in the ecliptic in the space of 9 hours for that day. The quantity of the diurnal motion for any day is found by taking the difference between it and the preceding day.

Thus let the diurnal motion for the 11th of May, 1787, be required.

	SIGNS.	DEG.	MIN.
On the 11th of May her place was	11	2	35
On the 10th of May	10	19	47
The diurnal motion sought		12	48

218   DESCRIPTION AND USE

TABLES

FOR FINDING THE HOURLY MOTION OF THE MOON, AND THEREBY HER TRUE PLACE AT ANY TIME OF THE DAY.

TABLE I.

HOURS	11 46	11 56	12 6	12 16	12 26	12 36	12 46	12 56	13 6	13 16	13 26
	d. m.	d. m.	d. m.	d. m.	d. m.	d. m.	d. m.	d. m.	d. m.	d. m.	d. m.
1	0 29	0 30	0 30	0 30	0 31	0 31	0 32	0 32	0 33	0 33	0 34
2	0 59	1 0	1 0	1 1	1 2	1 33	1 4	1 5	1 5	1 6	1 43
3	1 28	1 20	1 31	1 32	1 33	1 35	1 36	1 37	1 38	1 39	1 41
4	1 58	1 59	2 1	2 3	2 4	2 6	2 8	2 9	2 11	2 13	2 14
5	2 27	2 29	3 31	2 34	2 35	2 37	2 40	2 42	2 44	2 46	2 48
6	2 57	2 59	3 1	3 4	3 6	3 9	3 11	3 14	3 16	3 19	3 21
7	3 26	3 29	3 32	3 35	3 38	3 40	3 43	3 46	3 49	3 52	3 55
8	3 55	3 59	4 2	4 6	4 9	4 12	4 15	4 19	4 22	4 25	4 20
9	4 25	4 28	4 32	4 36	4 40	4 43	4 47	4 51	4 55	4 58	5 2
10	4 54	4 58	5 3	5 7	5 1	5 1	5 19	5 23	5 27	5 32	5 56
11	5 24	5 28	5 33	5 37	5 42	5 4	5 51	5 56	6 0	6 3	6 9
12	5 53	5 53	6 3	6 8	6 13	6 18	6 23	6 28	6 33	6 38	6 43
13	6 22	6 28	6 33	6 39	6 44	6 49	6 55	7 0	7 6	7 11	7 17
14	6 52	6 58	7 3	7 9	7 15	7 21	7 27	7 33	7 38	7 44	7·50
15	7 21	7 27	7 34	7 40	7 46	7 52	7 59	8 5	8 11	8 17	8 24
16	7 51	7 57	8 4	8 11	8 17	8 24	8 31	8 37	8 44	8 51	8 57
17	8 20	8 27	8 34	8 41	8 48	8 55	9 3	9 10	9 17	9 24	9 31
18	8 49	8 57	9 4	9 12	9 19	9 27	9 34	9 42	9 49	9 57	10 4
19	9 19	9 26	9 35	9 43	9 51	9 58	10 6	10 14	10 22	10 30	10 38
20	9 48	9 56	10 5	10 13	10 21	10 30	10 38	10 47	10 55	11 3	11 12
21	10 17	10 26	10 36	10 44	10 53	11 1	11 10	11 19	11 27	11 36	11 43
22	10 47	10 56	11 6	11 15	11 21	11 33	11 42	11 51	12 0	12 10	12 19
23	11 17	11 25	11 36	11 46	11 55	12 4	12 14	12 24	12 33	12 43	12 52
24	11 46	11 56	12 6	12 16	12 26	12 36	12 46	12 56	13 6	13 16	13 26

OF THE GLOBES.

TABLE II.

HOURS.	13 30	13 40	13 50	14 6	14 10	14 26	14 36	14 46	14 56	15 6	15 16
	d. m.	d. m.	d. m.	d. m.	d. m.	d. m.	d. m.	d. m.	d. m.	d. m.	d. m.
1	0 31	0 34	0 35	0 36	0 36	0 36	0 36	0 37	0 37	0 38	0 38
2	1 8	1 9	1 16	1 10	1 11	1 12	1 13	1 14	1 15	1 15	1 16
3	1 42	1 42	1 46	1 46	1 47	1 48	1 49	1 51	1 51	1 53	1 54
4	2 16	2 8	2 19	2 21	2 22	2 24	2 26	2 28	2 20	2 31	2 33
5	2 50	2 52	2 54	2 56	2 58	3 0	3 3	3 5	3 7	3 9	3 11
6	3 24	3 26	3 29	3 31	3 34	3 39	3 39	3 41	3 45	3 46	3 9
7	3 58	4 1	4 4	4 7	4 10	4 10	4 15	4 18	4 21	4 24	4 7
8	4 32	4 35	4 39	4 42	4 45	4 49	4 52	4 55	4 59	5 2	5 5
9	5 6	5 10	5 13	5 17	4 21	5 25	5 28	5 32	5 36	5 40	5 43
10	5 40	5 42	5 48	5 52	5 57	6 1	6 5	6 9	6 13	6 17	6 22
11	6 14	6 19	6 23	6 28	6 32	6 37	6 41	6 46	6 51	6 55	7 0
12	6 48	6 53	6 50	7 3	7 8	7 13	7 28	7 23	7 28	7 33	7 28
13	7 22	7 27	7 33	7 38	7 44	7 49	7 54	8 6	8 5	8 11	8 10
14	7 56	8 0	8 8	8 13	8 19	8 25	8 31	8 37	8 43	8 48	8 54
15	8 30	8 36	8 42	8 49	8 55	9 1	9 7	9 14	9 20	9 26	9 32
16	9 4	9 11	9 17	9 21	9 12	9 37	9 44	9 51	9 57	10 4	10 11
17	9 38	9 45	9 52	9 59	10 20	10 13	10 20	10 28	10 33	10 42	10 49
18	10 12	10 19	10 27	10 34	10 42	10 49	10 57	11 4	11 12	11 19	11 27
19	10 46	10 54	11 5	11 10	11 18	11 26	11 34	11 41	11 49	11 57	12 5
20	11 29	11 38	11 37	11 24	11 8	12 2	12 10	12 18	12 17	12 35	12 42
21	11 58	12 3	12 11	12 20	12 9	12 38	12 40	12 55	13 4	13 13	13 21
22	12 28	12 37	12 46	12 55	13 5	13 14	13 23	13 33	13 41	13 50	13 50
23	13 2	13 12	13 21	13 31	13 43	13 59	13 59	14 9	14 10	14 28	14 38
24	13 36	13 46	13 56	14 6	14 16	14 26	14 36	14 46	14 56	15 6	15 16

The moon's path may be represented on th[e] globe in a very pleasing manner, by tying [a] silken line over the surface of the globe exact[ly] on the ecliptic; then finding, by an ephemeri[s] the place of the nodes for the given time, co[n]fine the silk at these two points, and at 90 d[e]grees distance from them elevate the line abo[ve] 5¼ deg. from the ecliptic, and depress it [as] much on the other, and it will then represent t[he] lunar orbit for that day.

PROBLEM XLI.

To find the moon's place in the ecliptic, for any giv[en] hour of the day.

First without an ephemeris, only knowing t[he] age of the moon, which may be obtained fr[om] every common almanack.

Elevate the north pole of the celestial glo[be] to 90 degrees, and then the equator will be the plane of, and coincide with the broad pa[per] circle; bring the first point of Aries, marked on the globe, to the day of the new moon [on] the said broad paper circle, which answers to [the] sun's place for that day; and the day of [the] moon's age will stand against the sign and deg[ree] of the moon's mean place; to which place ap[ply] a small patch to represent the moon.

But if you are provided with an ephemeris,* that will give the moon's latitude and place in the ecliptic; first note her place in the ecliptic upon the globe, and then counting so many degrees amongst the parallels in the zodiac, either above or below the ecliptic, as her latitude is north or south upon the given day, and that will be the point which represents the true place of the moon for that time, to which apply the artificial sun, or a small patch.

Thus on the 11th of May, 1787, she was at noon in 2 deg. 35 min. of Pisces, and her latitude was 4 deg. 18 min.; but as her diurnal motion for that day is 12 48 in nine hours, she will have passed over 4 deg. 47 min. which added to her place at noon, gives 7 h. 22 min. for her place on the 11th of May, at nine at night.

PROBLEM XLII.

To find the moon's declination for any given day or hour.

The place in her orbit being found, by prob. xli, bring it to the brazen meridian; then the arch of the meridian contained between it and the equinoctial, will be the declination sought.

* The nautical almanack is the best English ephemeris.

PROBLEM XLIII.

To find the moon's greatest and least meridian altitudes in any given latitude, that of London for example.

It is evident, this can happen only when the ascending node of the moon is in the vernal equinox; for then her greatest meridian altitude will be 5 deg. greater than that of the sun, and therefore about 67 deg.; also her least meridian altitude will be 5 deg. less than that of the sun, and therefore only 10 deg.: there will therefore be 57 deg. difference in the meridian altitude of the moon; whereas that of the sun is but 47 deg.

N. B. When the same ascending node is in the autumnal equinox, then will her meridian altitude differ by only 37 deg.; but this phenomenon can separately happen but once in the revolution of a node, or once in the space of nineteen years: and it will be a pleasant entertainment to place the silken line to cross the ecliptic in the equinoctial points alternately; for then the reason will more evidently appear, why you observe the moon sometimes within 23 deg. of our zenith, and at other times not more than 10 deg. above the horizon, when she is full south.

PROBLEM XLIV.

To illustrate, by the globe, the phenomenon of the harvest moon.

About the time of the autumnal equinox, when the moon is at or near the full, she is observed to rise almost at the same time for several nights together; and this phenomenon is called the *harvest moon.*

This circumstance, with which farmers were better acquainted than astronomers, till within these few years, they gratefully ascribed to the goodness of God, not doubting that he had ordered it on purpose to give them an immediate supply of moon-light after sun-set, for their greater convenience in reaping the fruits of the earth.

In this instance of the harvest moon, as in many others discoverable by astronomy, the wisdom and beneficence of the Deity is conspicuous, who really so ordered the course of the moon, as to bestow more or less light on all parts of the earth, as their several circumstances or seasons render it more or less serviceable.*

About the equator, where there is no variety of seasons, moon-light is not necessary for gathering in the produce of the ground; and

* Ferguson's Astronomy.

there the moon rises about 50 minutes later every day or night than on the former. At considerable distances from the equator, where the weather and seasons are more uncertain, the autumnal full moons rise at sun-set from the first to the third quarter. At the poles, where the sun is for half a year absent, the winter full moons shine constantly without setting, from the first to the third quarter.

But this observation is still further confirmed, when we consider that this appearance is only peculiar with respect to the full moon, from which only the farmer can derive any advantage; for in every other month, as well as the three autumnal ones, the moon, for several days together, will vary the time of it's rising very little; but then in the autumnal months this happens about the time when the moon is at the full; in the vernal months, about the time of new moon; in the winter months, about the time of the first quarter; and in the summer months, about the time of the last quarter.

These phenomena depend upon the different angles made by the horizon, and different parts of the moon's orbit, and that the moon can be full but once or twice in a year, in those parts of her orbit which rise with the least angles.

The moon's motion is so nearly in the

ecliptic, that we may consider her at present as moving in it.

The different parts of the ecliptic, on account of it's obliquity to the earth's axis, make very different angles with the horizon as they rise or set. Those parts, or signs, which rise with the smallest angles, set with the greatest, and *vice versa*. In equal times, whenever this angle is least, a greater portion of the ecliptic rises, than when the angle is larger.

This may be seen by elevating the globe to any considerable latitude, and then turning it round it's axis in the horizon.

When the moon, therefore, is in those signs which rise or set with the smallest angles, she will rise or set with the least difference of time; and with the greatest difference in those signs which rise or set with the greatest angles.

Thus in the latitude of London, at the time of the vernal equinox, when the sun is setting in the western part of the horizon, the ecliptic then makes an angle of 62 deg. with the horizon; but when the sun is in the autumnal equinox, and setting in the same western part of the horizon, the ecliptic makes an angle but of 15 deg. with the horizon; all which is evident by a bare inspection of the globe only.

Again, according to the greater or less inclination of the ecliptic to the horizon, so a greater or less degree of motion of the globe about it's axis will be necessary to cause the same arch of the ecliptic to pass through the horizon; and consequently the time of it's passage will be greater or less, in the same proportion; but this will be best illustrated by an example.

Therefore, suppose the sun in the vernal equinox, rectify the globe for the latitude of London, and the place of the sun; then bring the vernal equinox, or sun's place, to the western edge of the horizon, and the hour index will point precisely to VI; at which time, we will also suppose the moon to be in the autumnal equinox, and consequently at full, and rising exactly at the time of sun-set.

But on the following day, the sun, being advanced scarcely one degree in the ecliptic, will set again very nearly at the same time as before; but the moon will, at a mean rate, in the space of one day, pass over 13 deg. in her orbit; and therefore, when the sun sets in the evening after the equinox, the moon will be below the horizon, and the globe must be turned about till 13 deg. of Libra come up to the edge of the horizon, and then the index will point to 7 h. 16 min. the time of the moon's rising, which is an hour and quarter after sunset for dark night. The next day following there will be $2\frac{1}{2}$ hours, and so on successively, with an increase of $1\frac{1}{4}$ hour dark night each evening respectively, at this season of the year; all owing to the very great angle which the ecliptic makes with the horizon at the time of the moon's rising.

On the other hand, suppose the sun in the autumnal equinox, or beginning of Libra, and the moon opposite to it in the vernal equinox, then the globe (rectified as before) being turned about till the sun's place comes to the western edge of the horizon, the index will point to VI, for the time of the setting, and the rising of the full moon on that equinoctial day. On the following day, the sun will set nearly at the same time; but the moon being advanced (in the 24 hours) 13 deg. in the ecliptic, the globe must be turned about till that arch of the ecliptic shall ascend the horizon, which motion of the globe will be very little, as the ecliptic now makes so small an angle with the horizon, as is evident by the index, which now points to VI h. 17. min. for the time of the moon's rising on the second day, which is about a quarter of an hour after sun-set. The third day, the moon will rise within half an hour; on the fourth, within three quarters of an hour, and so on; so that it will be near a week before the nights will be an hour without illumination; and in greater latitudes this difference will be still greater, as you will easily find by varying the case, in the practice of this celebrated problem, on the globe.

This phenomenon varies in different years; the moon's orbit being inclined to the ecliptic about five degrees, and the line of the nodes continually moving retrograde, the inclination of her orbit to the equator will be greater at some seasons than it is at others, which prevents her hastening to the northward, or descending southward, in each revolution, with an equal pace.

PROBLEM. XLV.

To find what azimuth the moon is upon at any place when it is flood, or high water; and thence the high tide for any day of the moon's age at the same place.

Having observed the hour and minute of high water, about the time of new or full moon, rectify the globe to the latitude and sun's place; find the moon's place and latitude in an ephemeris, to which set the artificial moon,* and screw the quadrant of altitude in the zenith; turn the globe till the horary index points to the time of flood, and lay the quadrant over the center of the artificial moon, and it will cut the horizon in the point of the compass upon

---
* Or patch representing the moon.

OF THE GLOBES. 229

which the moon was, and the degrees on the horizon contained between the strong brass meridian and the quadrant, will be the moon's azimuth from the south.

To find the time of high water at the same place.

Rectify the globe to the latitude and zenith find the moon's place by an ephemeris for the given day of her age, or day of the month, and set the artificial moon to that place in the zodiac; put the quadrant of altitude to the azimuth before found, and turn the globe till the artificial moon is under it's graduated edge, and the horary index will point to the time of the day on which it will be high water.

THE USE OF THE CELESTIAL GLOBE IN THE SOLUTION OF PROBLEMS ASCERTAINING THE PLACES AND VISIBLE MOTIONS OF ORBITS OR COMETS.*

There is another class or species of planets which are called *comets*. These move round the sun in regular and stated periods of times in the same manner, and from the same cause as the rest of the planets do; that is, by a centripetal force, every where decreasing as the

* Martin's Description and Use of the Globes.

squares of the distances increase, which is [the] general law of the whole planetary system.

this centripetal force in the comets being c[om]pounded with the projectile force, in a very [dif]ferent ratio from that which is found in [the] planets, causes their orbits to be much mor[e el]liptical than those of the planets, which are [the] most circular.

But whatever may be the form of a com[et's] orbit in reality, their geocentric motions, [or] the apparent paths which they describe in [the] heavens among the fixed stars, will always [be] circular, and therefore may be shewn upon [the] surface of a celestial globe, as well as the [mo]tions and places of any of the rest of [the] planets.

To give an instance of the cometary pr[oblems] on the globe, we shall chuse that comet, for [the] subject of these problems, which made it's [ap]pearance at Boston, in New England, in [the] months of October and November, 1758, in [it's] return to the sun; after which, it approac[hed] so near the sun, as to set *heliacally*, or to be [lost] in it's beams for some time spent in pas[sing] the perihelion. Then afterwards emerg[ing] from the solar rays, it appeared retrograde [in] it's course from the sun towards the latter [part] of March, and so continued the whole mo[nth] of April, and part of May, in the West Ind[ies] dered it visible in those parts, when it was, for the greatest part 'of the time, invisible to us, by reason of it's southern course through the heavens.

When two observations can be made of a comet, it will be very easy to assign it's course, or mark it out upon the surface of the celestial globe. These, with regard to the above-mentioned comet, we have, and they are sufficient for our purpose in regard to the solution of cometary problems.

By an observation made at Jamaica on the 31st of March, 1759, at five o'clock in the morning, the comet's altitude was found to be 22 deg. 50 min. and it's azimuth 71 deg. south-east. From hence we shall find its place on the surface of the globe by the following problem.

PROBLEM XLVI.

To rectify the globe for the latitude of the place of observation in Jamaica, latitude 17 *deg.* 30 *min. and given day of the month, viz.* March 31*st.*

Elevate the north pole to 17 deg. 30 min. above the horizon, then fix the quadrant of altitude to the same degree in the meridian, or zenith point. Again, the sun's place for the 31st of March is in 10 deg. 34 min. ♈ which bring to the meridian, and set the hour index at XII, and the globe is then rectified for the place and time of observation.

PROBLEM XLVII.

To determine the place of a comet on the surface of the celestial globe from it's given altitude, azimuth, hour of the day, and latitude of the place.

The globe being rectified to the given latitude, and day of the month, turn it about towards the east, till the hour index points to the given time, viz. V o'clock in the morning; then bring the quadrant of altitude to intersect the horizon in 71 deg. the given azimuth in the south-east quarter; then, under 22 deg. 50 min. the given altitude, you will find the comet's place, where you may put a small patch to represent it.

PROBLEM XLVIII.

To find the latitude, longitude, declination, and right ascension of the comets.

In the circles of latitude contained in the zodiac, you will find the latitude of the comet to be about 30 deg. 30 min. from the ecliptic; the same circle of latitude reduces it's place to the ecliptic in 26 deg. 30 min. of ♒, which is it's longitude sought. Then bring the cometary parch to the brazen meridian, and it's declination will be shewn to be 9 deg. 15 min. south. At the same time, it's right aseension will be 227 deg. 30 min.

PROBLEM XLIX.

To shew the time of the comet's rising, southing, setting, and amplitude, for the day of the obserservation at Jamaica.

Bring the place of the comet into the eastern semicircle of the horizon, (the globe being rectified as directed) the index will point to III hours 15 min. which is the time of it's rising in the morning at Jamaica, the amplitude 10 deg. very nearly to the south. The patch being brought to the meridian, the index points to IX o'clock 10 min. for the time of culminating, or being south to them. Lastly, bring the patch to touch the western meridian, and the index will point to III in the afternoon, for the time of the comet's setting, with ten deg. of southern amplitude, of course.

PROBLEM L.

From the comet's place being given, to find the time of it's rising in the horizon of London, on the 31st day of March, 1759.

For this purpose, you need only rectify the globe for the given latitude of London, and bring the cometary patch to the eastern horizon, and the index points to III hours 45 min. for the time of it's rising at London, with about 14 deg. of south amplitude; then turn the patch to the western horizon, and the index points to II hours 25 minutes, the time of it's setting.

N. B. From hence it appears, the comet rose soon enough that morning to have been observed at London, had the heavens been clear, and the astronomers had been before-hand apprized of such a phenomenon.

PROBLEM LI.

To determine another place of the same comet, from an observation made at London on the 6th day of May, at ten in the evening.

On the 6th day of May, 1759, at ten at night, the place of the comet was observed, and it's distance measured with a micrometer, from two fixed stars marked μ and ν in the constellation called *Hydra*, and it's altitude was found to be 16 deg. and it's azimuth 37 deg. southwest; from whence it's place on the surface of the globe, is exactly determined, as in prob. xlvii. and having stuck a patch thereon, you will have the two places of the comet on the surface of the globe, for the two distant days and places of observation, as required.

PROBLEM LII.

From two given places of a comet, to assign it's apparent path among the fixed stars in the heavens.

The two places of the comet being determined by the observations on the 31st of March, 1758, and the 6th of May following, and denoted by two patches respectively, you mnst move the globe up and down, in the notches of the horizon, till such time as you bring both the patches to coincide with the horizon; then will the arch of the horizon between the two patches shew, upon the celestial globe, the apparent place of the comet in the interval between the two observations, and by drawing a line with a black lead pencil along by the frame of the horizon, it's path on the surface of the globe will be delineated, as required. And here it may be observed, that it's apparent path lay through the following southern constellations, viz. the tail of Capricorn, the tail of Piscis Australis, by the head of Indus, the neck and body of Pavo, through the neck of Apus, below Triangulum Australe, above Musca, by the lowermost of the Crosiers, across the hind legs and through the tail of Centaurus, from thence between the two stars in the back of the Hydra before-mentioned; after this, it passed on to Sextans Uraniæ, and then to the ecliptic near Cor Leonis, soon after which it totally disappeared.

PROBLEM LIII.

To estimate the apparent velocity of a comet, two places thereof being given by observation.

Let one place be ascertained near the beginning of it's appearance, and the other towards the end thereof; then bring these two places to the horizon, and count the number of degrees intersected between them, which being the space apparently described in a given time, will be the velocity required. Thus, in the case of the above-mentioned comet, you will find that it described more than 150 deg. in the space of 36 days, which is more than 4 deg. per day.

PROBLEM LIV.

To represent the general phenomena of the comet, for any given latitude.

Bring the visible path of the comet to coincide with the horizon, by which it was drawn, and then observe what degree of the meridian is in the north point of the horizon, which, in the case of the foregoing comet, will be the 23 deg. This will shew the greatest latitude in which the whole path can be visible in any latitude less than this, as that of Jamaica; where, for instance, the most southern part of the path will be elevated more than 5 deg. above the horizon, and the comet visible through the whole time of it's apparition. But rectifying the globe for the latitude of London, the path of the said comet will be for the most part invisible, or below the horizon; and therefore it could not have been seen in our latitude, but at times very near the beginning and end of it's appearance; because, by bringing the comet's path on one part to the south point of the horizon, it will immediately appear in what part the comet ceases to be visible; and then the bringing the other part of the path to the point, it will appear in what part it will again become visible.

After this manner may the problems rela[ting] to any other comets be performed; and the paths of the several comets, which [have] hitherto been observed, may be severally [deli]neated on the celestial globe, and their var[ious] phenomena in different latitudes be ther[eby] shewn.

www.ingramcontent.com/pod-product-compliance
Lightning Source LLC
Chambersburg PA
CBHW022057160426
43198CB00008B/259